"Death is not the greatest loss in life. The greatest
loss is what dies inside us while we live."

-Norman Cousins

ISBN: 978-0-577-15601-6

# Death of a Trillion Dreams

## Unknown Intelligent Forces and Their Genetic War on Humanity.

A memoir by former Skylab & Space Shuttle Systems Specialist,

## G. Donald Mitnik

This book is dedicated to:

Robert Crippen - Bruce McCandless - Owen Garriott

Story Musgrave - Joe Kerwin - Ed Gibson - Paul Weitz

Jerry Carr - Jack Lousma - Al Bean - Bill Pogue and...

Pete Conrad.

(R.I.P. my friend.)

...The astronauts of Skylab.

# Death of a Trillion Dreams

## Unknown Intelligent Forces and Their Genetic War on Humanity.

A memoir
by
G. Donald Mitnik

Due to the controversial nature of this book and for reasons of national security - Please be advised: The information you are about to read is entirely true. Your knowledge of these truths could jeopardize your future wellbeing in unknown ways. By reading any portion of this book, you will gain priority knowledge that certain government entities don't want you to know. The reader hereby exonerates the author, publisher and all who had a hand in the publication of this book, of all responsibility pertaining to any negative repercussions that may occur as a result of reading this book. In other words - proceed at your own risk.

### THIS IS NOT A JOKE.

# Introduction:

They will conquer the Earth but not a single shot will be fired. The human race, as we know it, will cease to exist and there's little we can do about it. The war started with the advent of radio and it may be another 300 years before it's over, but the end has already been determined. Their weapon of choice has taken lives by the thousands and has destroyed thousands of our most valuable assets. The attacks continually bombard us, yet we don't even know the enemy is at our door, and we won't ever know because worse than stealing our land, and worse than stealing our resources, this enemy, through genetic manipulations and viral infestations, is stealing the most precious thing we have, our very minds. Where are the Sir Isaac Newtons today? Where are the Albert Einsteins today? What has happened to this planet where, right now, one in every 150 births produces an autistic child and Alzheimer's disease hits one in every eight adults age 65 and older?

G. Donald Mitnik provides answers you may not want to hear but nevertheless are all too pragmatic to ignore. His insights are spellbinding and the reasons behind them are as compelling as it gets. Is there nothing we can do to save our beautiful Earth, this pearl of the galaxy from slipping through our fingers and into the clutches of those who will ravish and destroy all humanity? The answer may surprise you.

# About the Author:

    Throughout his 38-year career, working in the secure confines of the country's major aerospace companies, G. Donald Mitnik worked with scientists, engineers and astronauts on the highest priority security programs that the U.S. Department of Defense and NASA ever had on their dockets. As a senior spacecraft technician, in the early 1960s through the late 1970s he gained information on classified programs that to this day is still not discussed by those who were privileged enough to become part of the hand-picked crews that worked at elite levels within the world's most secure programs. Mitnik's informal conversations with the astronauts of Skylab, over a two-year period, led to the knowledge this book contains. Through a security oversight, one very well known astronaut told him of a thoroughly astounding experience he encountered while on an Apollo flight to the moon. Through this unique friendship with the astronauts, guards were let down and unbelievable events that had been hidden in government vaults for years were revealed. These highly classified accounts were divulged as true and Mitnik had to swear he would never reveal them to anyone. It was only after one astronaut's death, the 9/11 tragedy and 40 years of soul searching that Mitnik decided the world must know what the astronauts told him. These accounts have never been told to anyone else and their manifestations are destined to rock the very foundations of civilization.

# Table of Contents

# Chapter Four
**The nuclear bombs that will soon fall to Earth because of deteriorating orbits**

# Chapter Five
**The secret U.S. practice of releasing bio, germ and radiation contaminants on U.S. cities**

# Chapter Six
**The secret U.S. nuclear meltdown**

# Chapter Seven
**Reviewing data**

# Preface:

*Be sober, be vigilant; because your adversary the devil, as a roaring lion, walketh about seeking whom he may devour.*

*-1. Peter v.8.*

Everything in the universe is trying to kill you. If you were to travel just slightly outside of Earth's protective environment without a space suit you would be instantly dead. Your death would be due to a combination of negative pressure from the vacuum of space, the lack of breathable air in space, the extreme cold of space and the radiation that permeates space. These are all excruciatingly painful ways to die. Yes, the universe truly hates us. In fact, this tiny flake of rock called Earth is our only friend.

I'm a skeptic. I've never seen a UFO that could not be explained as an ordinary product of human or Earthly creation. However, I do realize this does not mean ETs aren't out there. In fact, given the colossal number of stars that are in the cosmos, most scientists would agree that intelligent life is certain to be found somewhere else in our galaxy. This leads to the now famous - Fermi Paradox; If so, where are they? I will endeavor to answer that question along with many other questions that have been raised concerning certain mysteries that, up until now, have had no real explanations that made any sense. In the following pages I will attempt to explain concepts that I know sound extraordinary at the onset. In the words of Carl Sagan, "Extraordinary claims require extraordinary proof." So, I will try to explain these topics with as much evidence as I can and I only ask that you, the reader, keep in mind that most of what I say in this book are not my own conclusions, but are instead the results of numerous casual and spontaneous conversations that took place in the early 1970s with the country's most respected astronauts, who at times followed their words with, "forget I said that" or "Remember, you didn't hear that from me." My sole purpose in writing these revelations now is to bring a public awareness to what I consider an extremely important time in human history. I feel that I must let everyone on the planet know serious problems exist for the human race and these problems may be impossible to resolve.

My career, before retiring, was as an aerospace Electrical/Electronics Technician. I planned and built products like the B-2 bomber, the F-22 fighter, the Space Shuttle and Skylab. I actually spent more time onboard the Space Shuttle and Skylab (when on the ground) than all of the astronauts put together - counting their time in space. At Cape Canaveral I ran wiring throughout the SIV-B stage of the Saturn V rocket that took men to the moon. I am very technically oriented and I have a keen need to know how all things work. In the past, I held government clearances that were considered above top secret. I'm of sound mind and body and, although I question everything, I don't consider myself to be a conspiracy theorist. I'm not psychotic or paranoid, regardless of what some might think of me for writing this. Much of what I'll reveal to you here all started back in 1971 when I was employed by McDonnell-Douglas Corporation in Huntington Beach, California as a Senior Space Vehicle Electrical Check-Out Technician (official title) on the Skylab Orbital Workshop. Skylab was NASA's predecessor to the International Space Station.

During pauses in procedural tests on the space vehicle, the company provided a well supplied double-wide mobile home that sat behind the main Vehicle Checkout Lab. The double-wide was for the rest and relaxation of astronauts and test crews. Some of the tests ran around the clock, so you can see the need for this otherwise extravagant coddling. Many of the astronauts whiled away their off time by playing cards or reading. Some just made small talk with the engineers and techs. There was a well-stocked refrigerator, plenty of pastries, coffee, sandwiches, soft drinks and an abundance of jovial camaraderie. It was in this relaxed environment that I heard things so astounding and extreme that I decided to jot them down after I got home at night. I later stored these notes away in a safe place. What I learned while in that double-wide was the most eye opening education one could ever hope for.

Today, the astronauts will certainly deny saying anything and I really can't blame them one bit. They are reminded, occasionally, about the security issues involved and a $10,000 fine - not to mention the loss of their salaries and generous pensions if they ever divulge what they swore to keep secret. No astronaut in his right mind will ever acknowledge what was said during Skylab's tests. I will not name all individual astronauts I spoke with as I don't wish to cause them any embarrassment. I can assure you, if I did tell you the names of those I spoke with, you would recognize some of them immediately. Not all of the astronauts who spent time at the company during this testing phase talked with me, in fact most did not. There were those who would pop in for one test and then quickly leave when it was completed. Among the more congenial astronauts were those who, to their credit, told nothing of their space flights or anything else concerning classified programs. But, right or wrong, there were those who talked very freely and it seemed as if no subjects were off limits. The astronauts were privy to an abundance of classified matters that were learned as part of necessary secret briefings given to them by agencies such as the CIA, NSA, and the Department of Defense. They needed to know all of what the "black departments" of the military knew concerning highly classified knowledge in order to deal with unexpected developments that could possibly take place during space flights. These briefings usually took place in clandestine underground vaults located in a variety of unassuming places throughout the country. Why did they discuss these classified accounts with me? There is a logical answer that goes beyond boredom or emotions.

As it turns out, among the many secret briefings the astronauts were given, there was one special Department of Defense meeting in 1967 concerning the (then) top secret military space program known as MOL - for Manned Orbital Laboratory. Most of the astronauts later assigned to Skylab and myself were given briefings on MOL and its top secret military operations. The program (which was later canceled) was actually a covert operation for two reasons. Number one: It was to be a polar orbiting spy satellite. Number two: It could be used as a manned base of operations for controlling or maintaining nuclear arsenals in space. The "black" (covert) side of the program was code named "Dorian" and it had all the clandestine architecture that was common for covert operations during that time. These protocols included, keypad operated double locks on vault doors leading to "black" facilities. You could not mention the names of contractors or sub-contractors, or even the cities where they were located. You were supplied with special coded badges that identified different levels of clearances and above all, there was an identity verification system between persons that used coded phrases. In the case of MOL a.k.a. "Dorian", no one could speak of any covert subjects to another person unless "introduced" by a person of mutual "Dorian" compliance known to both involved parties. If I, or anyone else needed to discuss a covert component of the MOL, with a person of unknown security status, a mutually "known" individual would need to be located to "introduce" us using the phrase, "…this is a formal Dorian introduction". We could, only then, discuss the classified part of the program – providing the classified level on their badge conformed to the protocol. This was the procedure used for "Dorian" as well as for an even more secure and highly classified operation concerning UFOs and aliens which was (at the time) code named "Majic".

Unlike "Dorian", I was not briefed on the "Majic" program and really only heard rumors about it. Well, as it happened, one astronaut - knowing I was cleared for the "Dorian" program somehow got things a bit skewed and "introduced" me to another astronaut using the term "formal Majic" instead of "formal Dorian". My jaw must have dropped ten inches but I said nothing of the blurb. It appeared he never realized his mistake (or if he did, chose never to admit it to anyone) and from that point on, things really got interesting as I was now treated by the astronauts as a person fully briefed for "Majic". This same astronaut later told me he realized his mistake but ignored the blunder, as instructed in briefings, because there is no way to backtrack (without making things worse) once a security slip is made. I assured him all was safe with me. He said "good" then added, "besides, no one - outside the covert program, would ever believe any of this stuff (only he didn't say stuff) anyway."

Several astronauts, who I worked with on Skylab, have since passed on. One astronaut in particular, who owed me many favors for getting the right test equipment to him during crucial tests, told me what he asserted to be a true account of a strange encounter he had while on board one of the Apollo flights. They say truth is stranger than fiction and in this case the truth, that he made me swear I'd never reveal, is so strange it's absolutely astounding. I'll tell you more about this later, but let me tell you right now, what is happening in outer space is well known by our military and the Department of Defense. However, very few people within the three branches of the federal government have a clue as to what the truth really is. I swore to my astronaut co-workers that I would never mention the conversations that took place during that period of Skylab's checkout, but as I'm not getting any younger and I feel the public has a right to know their own fate, through these pages I'll tell you the truth as it was told to me. Sometimes I'll draw my own conclusions or interpretations to accounts, and I invite you to do the same. The basic topics presented here remain honest, straight forward and politically unconstrained. Now let me tell you a little more about myself.

I retired from aerospace at a time when both my daughters were grown and pretty much on their own. Up until retirement, my wife and I almost always took our vacations in a kind of magical place called Sedona, Arizona. It's a beautiful red rock, high desert community about 100 miles north of Phoenix. We decided, when the time for retirement came, we would move there. So we did. I have always had a little bit of art talent and did some figurative sculpting even before retiring but once in Sedona, an art town to be sure, I started sculpting more and that has now become a kind of occupation. I've shown my work in several galleries and enjoy the relaxing outlet sculpting provides. In truth I am a person who has many interests - some of them conflicting. I love cars, from hot rods to European classics, yet I drive a minivan. Auto racing is also one of my passions. This stems from my days as a youngster when my Dad would take me to see the mighty midgets roar at the old Civic Stadium in Buffalo, N.Y. way back in the1940s. Yet, the only racing I've ever done is on my Playstation. I lived in Buffalo until I was nine years old - then moved, with my family, to Southern California where I became an avid comic book reader. I still have my collection of E.C. Comics, the ones that were banned in the '50s for being too deplorable for young minds. As I turned out to be (somewhat) normal, I choose to believe I'm living proof those indictments were completely wrong. I also like other odd things like reading about quantum physics and what the universe is about. I don't own a telescope but astronomy fascinates me. I become absorbed in archeology and history. Higher mathematics intrigues me - yet I have trouble figuring the change I should be getting at the supermarket. I take an interest in the ups and downs of the stock market - yet I don't own a single stock. Most of all, it's what I don't know that interests me the most. I find thinking about the unknown possibilities that interweave with the incredible knowledge I've been told by the astronauts, mind boggling and although I don't like to write all that much, I've found it impossible not to have written this book.

What is written here is not a theory. It is fact that involves every human on this planet and will have a profound effect on your future and the way you interpret life. You will notice this book does not have footnotes or references to other authors. Unless otherwise stated, the words here are entirely mine – grammatical errors and all. If any reader wants specifics on any given subject this book conjures up, they are invited to consult that great provider of truth in cyberspace called the Internet. I will not bore you with details or try to impress you with a lot of quotations and verbal tap dancing. One thing I have tried hard not to do is fluff up the book's word count in order to impress some literary agent or publishing house. It seems they think the amount of words used within a book is somehow related to the meaningful content. In their world, the thicker a book appears, the higher a price can be asked. In my world, the more concisely the information is presented, the more the reader benefits. There is absolutely no need to say the same things two or three times just to make a 40,000 word book into an 80,000 word book. I believe you - the reader - want what I have to report in the most efficient and least time consuming way possible and that is what I have strived to do. I have not created this book to pitch to an agent or publisher or to make big profits. Monetarily, I doubt if I'll break even. It is written strictly for you - so you can know the possibilities that lie in your future and can act accordingly. Everything written here is simply my own commentary on subjects that I learned from conversations in that double wide behind the Vehicle Checkout Lab in Huntington Beach some 40 years ago. Many have said I'm crazy to even bring them up, but I believe this story needs to be told. In the end, you are the judge. I know what I state is true. The evidence is there for all who have an open mind. I'll try to present it as best I can without prejudice or spin. I'll cover all the facts as honestly and as straightforward as I possibly can and even try to present counter points to many of the weirdest things you are about to hear.

As I've said, I'm a skeptic ...and I expect you to be one also.

---

# Chapter One
# The plan for human annihilation

## Part One: Manifest Destiny

From the time the first white settlers stepped onto The Americas, an attitude of superiority and indifference prevailed among them. First came the discoverers who claimed the land for their Monarchies. Then came the trappers, builders and farmers who divinely believed they deserved the riches of a land they knew little of. They came because of fear and greed. Fear from the lands they were leaving and greed for the riches a new land promised. They came with the consent of their Kings and who could question that? They also came to escape religious persecution. God was their guide and this was the New World he brought them to rein over. Never mind that it was a very old world to others.

So it began. For the next few hundred years their might and technology dominated all the native tribes that lived in this land we now call the United States. The strong overwhelming the weak has been the rule of human societies since the Neanderthal's demise some 30 million years ago. The Neanderthal's extinction was a direct result of the rise and dominance of modern man. Whatever point in history you prefer to examine, man's passions dictate that he must be the conqueror over other men. It has been tribe against tribe, village against village, state against state and nation against nation and so it will forever be. From the killing fields of Genghis Khan to the killing fields of Iraq and Afghanistan, war is our inherent predestination. When the Spanish conquistadors marched across Central and South America, they thought nothing of wiping out whole villages, women and children included, simply for the sake of finding gold. Gold to them was a precious resource that meant added riches for Spain and added glory for their cause. The real cause of most wars is either religion or the need for natural resources - be it gold, land, water or energy in the form of oil. Most of these commodities are rare but the Earth is a very good provider of the life-giving assets that we take for granted such as pure water, vast plains for growing food, a breathable atmosphere, and radiation protection in the form of a substantial magnetic field. This planet is truly a ripe peach ready to harvest.

When the Nazis invaded Poland, it was to take their natural and human resources. Enslaving one's enemy is also one of the many spoils of war. In war, humans can be treated as nonhuman without so much as a glimmer of conscience. Hitler's experiments on Jews are notoriously well documented. Heartless and inhuman treatment has been the case for countless other war prisoners and refugees throughout time. There are exceptions to this rule of ruthless behavior in war that seem to crop up for brief periods. The United States, as a country, seems to be going through an overall benevolent phase now - no torture of enemy combatants is an example. As nations and tribes pummel each other today, there is an attempt to spare civilian populations from destruction by using so-called smart bombs that pinpoint only military targets. The U.S. is also helping devastated governments re-form providing they re-form as we want them to re-form. This, of course, leads us to ask, why are we being so congenial? Could it be that we have learned that it's in our interest to have future access to their resources? I think it might be so.

What resources did the American Indians have that caused the white settlers to literally steal all their land within a 100-year period? The word land tells it all. Land to the Native American in the 18th and 19th centuries was, and still is, sacred. The Indian of the Americas in those days had no concept of land ownership. The land to them belonged to everyone and everyone had equal access to the land. Land was like the ocean, or the sky, or the stars. To the white men, however, land ownership was a right. To go west and homestead land was not only a right, but it was considered a noble and God given right. They called it Manifest Destiny and because they had the technology to crush any opposition, they, without conscience, prevailed. Indeed, with Manifest Destiny acting as a tailwind toward the west, their passionate nature told them, as with all conquerors, that God must truly be on their side and rolling over the "inferior" natives was the right thing to do. I'm not trying to denigrate religion here or even determine what was right and what was wrong. I'm just pointing out facts as I see them and perhaps my eyesight is different from yours. Let's say a lizard gobbles up a cricket, does that make the lizard evil? Maybe - in the eyes of the cricket - at least until the cricket devourers a fly. In the lizard, cricket, fly world - does good and evil even exist? Why should it exist in our realm or that of an alien species? With our human minds we perceive good and evil. Cruelties as in terrorism, harassment, war, torture and rape are all examples of man's evil tendencies. On the other hand we have our good aspects like charity, empathy, sympathy, creativity, art and science. Which all means, the religious principal of ..."do unto others..." can be very virtuous when dealing with everyday life, as it should be, but faith may take on a whole different meaning when a stronger and culturally different species comes into our domain. Whose side will God be on? In any case, the point I'm endeavoring to make is: Whenever a stronger, more technically advanced civilization meets a previously unknown and/or weaker civilization, chances are the former will prevail and pillage the later for whatever resources they deem valuable. This is our nature and as one of my astronaut co-workers put it, "It is surely in the nature of the highly advanced culture that is now in our midst." The reason for thinking this way has to do with the purpose a culture has for developing technology in the first place, and that basic underling purpose is a drive to be superior. This means any intelligence that has acquired technology will certainly have emotions, passions and the will to constantly press a way of life onto others to suit their purposes. In other words, any alien civilization out there with advanced technology will be emotionally similar to us. If this is the case, they will view earth as a treasure trove of resources just sitting here for their use. The only thing that might stand in their way is a small number of virtually weak beings that are pitifully corrupt and inept. These beings (us) however, do have the capability to mess things up a bit for them. The problem to them would be a minor one that they could resolve without these semi-intelligent beings (us again) ever knowing they've been had. It is a means of annihilation that leaves this gem of a planet totally intact without so much as a single shot or laser beam fired. It is clean, neat and effective. To their way of thinking they have every right to take over this planet. They are the rulers of the galaxy and this is their Manifest Destiny.

One other thing that should be of the utmost concern to anyone reading this: There have been radio transmissions emanating from Earth, at the speed of light, for approximately 100 years. These broadcasts, if picked up by alien beings, would tell them exactly who we are and where we live. In our innocence, we probably gave these conquerors a road map to our house. Now just in case radio alone didn't entice these friendly out of town folks to visit us, we've also had television and wireless broadcasting at a multitude of frequencies for more than half a century. Add to these invitations, our gall (in my opinion) at providing them with drawn maps and directions to Earth in the form of plaques put aboard space probes - Pioneers 10 and 11, two of our solar system leaving spacecraft, and recordings on board Voyagers 1 and 2 - all four probes launched in the 1970s - and you can count on visitors dropping in on us, whether we want them or not. In 1974, a booming message from the worlds' largest radio telescope at Arecibo, Puerto Rico was beamed into outer space telling anyone who happened to receive it exactly who we are and where we are located. All considered, this becomes somewhat like pre- Columbian American Indians taking out full page ads in 15th century editions of the Madrid and London newspapers, (if they had newspapers then) inviting all of the old world to a good old Native American powwow. I suppose most of these things couldn't be helped as the advances in mass media dictated it. But, there are definitely some cases here where discretion should have been applied. Please stay tuned for more of the same from our now famous galactic address. It's just a shame that the true consequences of these innocent "party invitations" - in all probability - will lead to our demise.

---

# Part Two: Who's Out There?

How do we know if anyone is out there? The answer to this lies in sheer numbers. The vast majority of scientists worldwide believe that life exists elsewhere in our galaxy. We'll stick with just The Milky Way Galaxy from here on because to think of travel between galaxies or life existing in other galaxies is just too much for my meager mind to contemplate. After all, the space frolicking folks in Star Trek never left The Milky Way galaxy and they had warp drive.

The chances of life popping up elsewhere in the realm of the galaxy is almost certain. This is the opinion derived by my astronaut friends and is also the conclusion brought forth in the famous "Drake Equation". The equation takes into account the number of stars in the Milky Way galaxy that possibly have stable rocky planets with breathable atmospheres. These planets must also have water and orbit their star in a temperate zone conducive to life. In addition, the planet's star must be stable and orbiting in a relatively quiet zone of the galaxy. These requirements may seem to indicate a rare chance of Earth-like planets existing but in fact the staggering number of stars in the galaxy makes the case for life, and intelligent, technologically advanced life in particular, almost absolute. If we are the galaxy's only intelligent life, that would truly be remarkable to the point of being unbelievably ludicrous. You might ask: When did nature make only one of anything? I can't think of an answer. Without a doubt, they are out there and what is more important; they are aware of us and are practicing their slow form of human genocide on us as you read this.

SETI, The Search for Extraterrestrial Intelligence has been ongoing for quite a number of years. The program basically searches the skies for radio transmissions at various frequencies in the hope of detecting a civilization that, like ours, is broadcasting using radio signals as a means of communication. There is also Optical SETI which searches for light signals such as laser transmissions that might be used by ETs. The consensus is, radio SETI probably should have made some sort of contact by now. Why haven't we at least heard some juicy alien gossip after all this time? Well, time itself may be the answer. We've been what you might call a technological society for only the better part of 150 years. That's hardly a blink in universal time. So it could be that our timing is off just a little. In a sense, it could be - they weren't home when we were listening, or perhaps we weren't home when they were talking. In all likelihood, I think they have a much more advanced means of communicating than we could ever imagine and to us it's simply undetectable...or, maybe they just don't talk to "bugs". I really believe this is why there has been no contact and probably never will be any contact. It's just the way they prefer it.

Within the last 20 years or so we've come up with some very slick advances in stealth technology. The F-117 and the B-2 bomber come to mind. Just try to imagine the stealth, or outright invisibility techniques of a civilization ahead of us by a mere 1,000 years, or 2,000 years, or 10,000 years. We don't see them or hear them unless they want us to. That brings us to the question of UFO sightings. We certainly hear enough about those. Worldwide, people from all walks of life, from airline pilots to police officers have claimed to have seen them. I don't think these people are crazy ...at least not all of them. I was told, what they are seeing are robotic craft that alien beings, from a place other than Earth, are manipulating to survey the Earth for natural resources and gather biological genetic information. For some reason, probably due to atmospheric conditions, these probes become visible to us at various times. A bigger issue lies in the fact that these UFOs can spread highly infectious viruses. This problem with viral contamination is the main reason for the government's continual UFO cover-up. A worldwide epidemic causing economic meltdown and panic is not what the government wants. In actuality, the defense department is trying to combat the threat of unearthly viruses by spreading anti-viral agents throughout the country via military and civilian aircraft. Everyone sees these "chem-trails", now you know what they are and the reason for them. Although somewhat effective on our own viruses, the spraying efforts will not stave off what has now become the inevitable onslaught of alien viruses that are slowly modifying the human genes that control brain functions. More on this later.

You might have heard reports that sightings of UFOs have increased substantially of late. The truth is exactly the opposite. What has increased are sightings of experimental aircraft along with a natural phenomenon brought about by atmospheric divergences and polar magnetic storms. Then you have outright hoaxes propagated by the government for the purpose of misleading the public. Mix in individuals with possible monetary gain in mind, continually claiming to have seen something that in reality does not exist and it's no wonder UFOs seem to be everywhere. The fact is, since the 1990s fewer legitimate sightings have taken place. This is exactly what you would expect to happen as these aliens continually improve their stealth and invisibility technology. My prediction is, there will be even fewer, "true" UFO sightings in the future. Anomalous observations might increase at times - mostly due to the hoaxes and government shenanigans - but, over the long term, real UFO sightings will decrease. This decrease will soften our outlook on UFOs, the subject will be ridiculed, laughed at - even more than it is today - and our vigilance will wane. Just the scenario an adversary would want.

Remember that Apollo, moon destined, astronaut back in the double-wide trailer at McDonnell-Douglas? His account, as he told it to me, was exactly as follows: Their Apollo flight was more than two thirds of the way to the moon when the astronauts were notified by Houston to switch to their emergency channel. This channel operates on an encrypted frequency and is used only during unusual or unpredictable circumstances. They switched channels and were told to take a peek out their view port to see if there was anything unusual out there. It seems an amateur astronomer somewhere in New Mexico had spotted the Apollo spacecraft with a small telescope and notified NASA of his sighting. NASA knew this was impossible because even the most powerful telescopes could not pick out Apollo's image at that distance and NASA's tracking equipment showed nothing in the vicinity of the spacecraft. At the time, the S-IVB stage that put them into a trans-lunar trajectory was verified as being six thousand miles from their position. To be on the safe side, as instructed, they looked outside. They could not believe their eyes. Moving in their direction, at a fairly high rate of speed, was an asteroid that appeared to be the size of the Titanic. They immediately told Flight Control what they saw and just as they were about to take evasive action, the asteroid inexplicitly and ever so slightly changed course. It missed them by about the length of a couple of football fields. There was nothing but complete silence for about thirty seconds. Then Houston came back to them with a simple, "Apollo, we've been advised... all parameters appear nominal." The astronauts were then told to not bring it up at the post flight briefing. They were reminded about the security clearances they signed and told not to talk to anyone concerning the incident because national security was at stake. Nothing was ever mentioned of the situation again. Even though these astronauts were told beforehand this situation could occur, it was unbelievably frightening when it actually happened.

I have been told by the astronauts of Skylab that aliens (extraterrestrials, for lack of a better description) are controlling the robotic probes that we occasionally spot and call UFOs. The aliens, themselves, are monitoring their operations from larger ships that have been observing us now for decades. These alien space vehicles are gigantic worlds by themselves and are in orbits that mimic asteroids. They not only stay stealthy by taking on the orbits of asteroids, they are, in all probability, actual asteroids. This is done not so much for our benefit, as their stealth techniques alone could no doubt keep us from detecting them. It is done to deceive other advanced beings who are also out there. It would be a simple process for them to take an existing asteroid and convert it for their use. What better way to travel through the galaxy, undetected, than on the surface, or tunneled into the depths of a giant asteroid that for short durations can appear invisible to us on Earth? As they drift lazily by our precious Earth, we mistakenly think an asteroid has passed by us at an astonishingly short distance and how lucky we are to have once again dodged the bullet. The truth is, these close calls will probably never hit us because they are being subtlety controlled by the aliens who inhabit them and the last thing they want is anything very big to collide with Earth. The last large asteroid (or possibly, comet) collision with Earth was in 1908 in remote central Siberia. The Tunguska "event" as it's called, although extremely large, happened in such a remote area that it wasn't discovered until 19 years later. Our alien observers knew it would have little effect on the Earth so they did nothing to divert it - if they could. This means another "Tunguska" could possibly hit us, but probably not in a populated area and with little effect on the environment. Have you ever questioned why some asteroids suddenly appear seemingly out of nowhere? I was told they are controlled and their trajectories as well as their "cloaking" abilities are altered at will as they "fly" into and out of Earth's view. It is a little known fact that gravitational tugs influence the trajectories of all asteroids and their paths are continually changing. This is the reason astronomers keep watch on known asteroids and update their trajectories as they are changed by various gravitational influences. The astronauts believed the aliens are traveling aboard genuine asteroids and not comets. This is possibly because the radiated and volatile sublimating gases inherent with comets would cause our traveling conquistadors many unforeseen problems. Only elite elements of the military and some "black" departments of NASA know the aliens are aboard certain asteroids. This is why NASA, along with other countries like Japan and China - who have been told only of non secure, scientific aspects related to the asteroids - are now deploying probes to these bodies in increasing numbers. There are departments within NASA that have known about these "controlled" asteroids for years and have actually debated the pros and cons of sending an Earth envoy consisting of a team of astronauts to (get this) greet them. At the time, I was told by Skylab's astronauts, we lacked some of the technology to accomplish such a mission, but the astronauts I spoke with did seem to favor this approach - in lieu of any other alternative - if it ever became possible in the future. If a trip like this does become reality, it would be interesting to see if the "diplomatic" crew ever comes back.

The alien occupied asteroids meander by our domain while in the process of commanding their stealthy smaller, robot ships down to Earth's surface to monitor how their diabolical plan for us is taking shape. Asteroid-based navigation of the galaxy certainly would have economic value over building an intergalactic vessel from scratch and would save on the time and talent it would take their robots to create such a ship. I have been told, most all of their labor is done by robots using a form of artificial intelligence they have programmed into them. We have been using robotic space probes to explore the planets and moons of our solar system for years. We also have remote-controlled drones that are used for aerial military purposes. They are using these same techniques, only in a much more sophisticated way. These robots do all the planning and implementation that will eventually bring about the results the aliens are impassioned to achieve. This is one of the main differences between the robots and the aliens themselves. The aliens have emotions and passion. The robots have no emotions whatsoever. This is their strength as they are free to make pragmatic decisions with impunity. When I use the term, "robot" I do not necessarily mean a robot with a head, two arms and two legs. Each robot is designed with a specific purpose in mind and perhaps the probes that we do see once in a while are not vehicles that are being flown by humanoid looking robots, or what we would call androids, but the craft itself is the robot. So their robots could have many forms. The aliens may even need the robots to make other robots since the aliens might not have the necessary dexterity or articulation to physically create what their technically oriented brains inspire them to create. We cannot begin to imagine what these aliens look like, or if we can even see them at all.

UFO sightings have been reported for at least a century and probably long before that. We all know of these reports and have seen the films and the documentaries numerous times. UFOs by definition are of unknown origin and none can be proved as being extraterrestrial in nature but, from what I was told in the strictest confidence, while most UFO sightings are explainable earthly phenomena, many are extraterrestrial in origin. Without a doubt the United States military has a program that uses new propulsion techniques and other advanced technologies on a large triangular craft that has been spotted throughout the country. They occasionally fly it over populated areas to stir reaction. This vehicle has been mistaken by some to be an alien UFO. In truth, it is part of the military's misinformation program to cover, confuse and condemn real UFO sightings. The point is, many reputable persons, from police officers to astronauts have seen, with their own conservative and rational eyes, things in the skies that are unexplainable. These are all people who certainly know the difference between an experimental aircraft, clouds, the planet Venus and actual sightings of what are truly unexplained phenomena.

During my days in aerospace I can remember casual conversations with many of the astronauts assigned to Skylab. They told of numerous other unexplained sightings that they experienced while on space flights. They believed these sightings were of the alien's robot craft and these robots (more probably androids) are now doing things that will make your hair stand on end. I was told they are causing the most complete human extinction to ever hit our planet and they are doing it essentially undetected. They don't want us to know a thing and that is exactly how their plan is taking shape. They simply want to destroy humanity while keeping the Earth and all its natural resources intact. Except for getting rid of us, their aim is to not touch any of the fragile animal or plant life that makes this planet so rich. By eliminating humans they will eliminate the one element that is polluting the water, changing the atmosphere, spreading viruses throughout the world, and through wars, threatening to radiate and possibly blow up the entire Earth. Is this reason enough for them to want us out of the picture? Unfortunately for us, the answer is a resounding yes.

Who's out there? The answer is everybody and everything you can imagine. They, of course, in their minds, might see themselves as Gods, working for the greater good of the entire galaxy and perhaps they are. Their role is to be the conquistadors and we are stuck being the Aztecs. Is it just me, or does doom for the entire human race seem like a bad idea?

---

# Part Three: A Diamond in the Sky

If we had the ability to find and travel to another planet like the Earth, what do you suppose we would do with all those extra natural resources? What would we do especially if the most advanced animals on that planet were...let's say, cockroaches?

Mining the galaxy, that's their business. Whatever they're looking for, be it gold, diamonds, copper, carbon dioxide, oxygen, uranium or something we can't imagine, chances are it's here, and here in abundance. Earth, with all its wonderful riches waits for them. If someone comes to steal your possessions from your home, at least you have some protection from the police. You might even have safeguards like an alarm system. But who are we going to call when they come to steal the Earth? Most of the government, your congressmen and senators, do not have the foggiest idea that aliens are stalking us and all the military can do is shoot ineffective lasers at them - which they seem to easily avoid. The public, in reality, has no idea of the divide that exists between the government and the military concerning UFOs. The government is kept in the dark while the military has a secret program that spreads UFO absurdities just to redirect public interest away from true classified programs. The government is left holding a bag full of denials while the Department of Defense keeps what it knows a secret more tightly guarded than the atom bomb was during WWII. Most Presidents have been unaware of the aliens and the threat they impose. The last President to be informed was Ronald Reagan, who almost blew it with his famous "alien threat" speech and thus will probably remain the last President to be told. Reagan, in an effort to bring the Russians and U.S. closer, consequently told Mikhail Gorbachev, - yes, the Russians know. Sharing this knowledge with the (then) U.S.S.R. was what helped lead to an era of understanding and relative cooperation between the U.S. and the Soviets. It may have even lead to the downfall of the Soviet Union and a new emergence of democracy in most of the former Soviet bloc nations. The Russian government, like the United States, has decided to keep what it knows about the alien situation a top security priority because (and again like the United States) there is simply nothing anyone or any government can do about it. Neither government will ever admit that they cannot defend their people. They believe chaos and economic disaster would be the end result and numerous studies show they are right. For now, both nations are content to let the status quo prevail and misinformation abound. The result of U.S. concocted misinformation, for most of the world's unknowing populous, has been confusion that the military revels in. Secret research programs, ardent denials, and misinformation programs are all going on at the same time. It is this inept government mismanagement that may have contributed to the alien discovery of us in the first place. I know the following sounds crazy, but it is nonetheless true. Forty five years ago the United

States engaged in a program that, except for the releasing of germ toxins on U.S. and Canadian cities (more on this later), was the most unbelievably reckless series of tests ever devised by a government. With the end results being totally unknown, the U.S. proceeded to explode a series of nuclear bombs up in the Earth's magnetic field just to see what the affect would produce. To this day, the total results are still classified as top secret. What is known is that significant damage was done to our planet's ability to protect us from radiation and this damage is not repairable. The point I'm trying to make here is; past government involvements in programs like this tend to attract alien interest in taking the Earth's natural resources away from us and justifiably putting an end to our meddling. (I'll tell you more of this nuclear atrocity later).

I have been told that the aliens need our natural resources to convert into energy. Energy to run their vessels, energy for their robotic probes and quite possibly, energy to just do their laundry. So, a very good reason they would have to take possession of our domain would be energy. Secondly, they might need breathable air and drinkable water. I have also been told, by the astronauts, they might need the Earth to use as a giant toilet. A dump if you prefer, to abandon their wastes. Although I personally don't think this is the case, since they seem to want to keep our world in pristine condition. The one, overwhelming reason they are here is: Like the conquistadors, they are searching for El Dorado, the lost city, or in this case - planet, of gold or something else that happens to be valuable and rare to them. Whatever this commodity is, to us it could very well be excessive and relatively worthless. You might want to fill in the blank here since your guess is as good as mine (or the astronauts) as to what this could be. It could be anything. Pine tar or Limburger cheese or maybe a box of dirt to take back to their gas planet's King. In any event, Earth is a gem that they have "discovered" and it is now their means to fulfill their galactic plans. How rare it must be to reach a jewel of a planet like this - a single gleaming marble in a sea of vast darkness. The promise it provided when they first laid eyes on it must have been overwhelming. Eureka! They hit the lotto, the jackpot of all jackpots. My guess is, this is not the first time they've come across a hapless "village" like the Earth and I'll also guess they know just what to do with us.

In a sense, if they were only after our resources, Earth to them would be nothing more than a pit mine was to a '49er. On the other hand, if they intended to live here, they would treat the planet with some respect. Since the later seems to be the case, I believe they plan to keep every single part of this planet as it is, as they prepare for an eventual occupation. Let's just say it's oil they're after and the Earth is the only source they have come across in their journey through the galaxy. They might not even need the oil for themselves, since they more than likely have energy sources we can't begin to imagine. They would probably use our oil to trade or bargain with other entities out there that do need it. So whatever they seek in the way of our natural resources, it might not be for their own use. Our Earth produces commodities like oil in rare abundance. Uranium is a good candidate for their desires, since the use of nuclear energy could be one of the ways they produce the power it takes to keep their civilization controlled and on the move. They may even process it aboard their asteroid ships or store it there. They could store most anything they want on their colossal "freighters" because the cores of these asteroids would provide built-in shielding from radiation. This radiation shielding could work two ways. It could shield them from their own manufactured radiation and from the radiation in space that is generated by the stars.

Another candidate for their interest in Earth is our oceans. Water in great amounts could be needed by them in order to replenish their dwindling supplies. Their own water may have become contaminated. The aliens would definitely need a supply of water on board their asteroid ships and what better place is there to obtain water than Earth with two thirds of it covered with the stuff. Many times UFOs have been reported diving into our seas and blasting out of our seas. Could these be those pesky robot ships exploring or testing or even stealing our water?

Salt is another commodity they might be after. Salt has always been in demand here, why not throughout the galaxy? Maybe they're after the biological creatures that permeate our seas, and our land for that matter. They could yield benefits for medical use, drugs and medicines of unimagined kinds. The cornucopia that is our Earth might, with the technology the aliens have, be the galaxy's one and only source for guaranteeing their very existence. For sure, there's something here they want, perhaps all the above and more, and they'll stop at nothing to get it.

The Earth is truly the "Goldilocks" planet. Not too hot and not too cold, it's just right. At least for us and, more than likely, for them as well. Tilted at about 24 degrees from the plain of its orbit, we have distinctive seasons that stabilize our worldwide weather patterns. Our moon, a small planet by any standard, orbits Earth at close range. Its gravitational pull creates the tidal forces needed to balance and equalize plate tectonics which accounts for the formation of our continents. How rare indeed we must be, especially to a civilization that might not have a home planet. Their home planet could have met its demise eons ago. They could have used up all their natural resources, or simply outgrew their world. They may have destroyed their home through wars or some other annihilating factor. Collisions with other orbiting bodies could have wiped them out or it might have been the death of their star. In any event, they may have been searching for just the right place to call their own for hundreds of generations and now, having found it, they are not about to let anything stop them from calling Earth their new home. They would have to get accustomed to 24 hour days, and they might have to get accustomed to a bit more or a bit less gravity, and they might even have to get use to some new foods. Oh yes, new foods. Think about that one for a minute. What could there possibly be here that they might eat? Although the answer could be us, I don't think it is. With our abundance of plants and cattle, if they're meat eaters, I tend to think we'll not become their next soufflé. I really think through years of interstellar travel, they have developed their own artificial foods, atomic proteins, for lack of a better term, and humans as food is the last thing they need from the Earth. I admit, I could be wrong here but I don't think so. Which leads to the question, what do they really want? Again, they want whatever their form of "gold" happens to be. Would we negotiate and give it to them, in trade for their technology or perhaps their word that they will not be hostile toward us? Of course we would, but how good is their word when dealing with us would be like us dealing with the cockroaches I mentioned before. It's just simpler and less messy to have us out of the way.

# Part Four: The Human Problem

Ah, the many ways to get rid of us. What would be the most convenient for them? Let's examine the possibilities and count the ways. First, they surely would have numerous methods for annihilating their adversaries. Given their position in space, the high ground so to speak, they could possibly cause large rocks or asteroids to rain down on us. We all know what just one of these things did to the dinosaurs. How about tilting the Earth's axis just a little? This surly would send us into a frenzy if not oblivion. Then there's always the way it was done in the movies, like blasting us with death rays or completely destroying our cities with some super high-powered bombardment of an unknown energy force that turns us all into dust. Well, why not? Although these measures would certainly do away with us, they, at the same time, cause many other problems for the aliens. Their goal is to keep this immaculate and delicately balanced planet as it is. That means no burning, no radiating and no blasting of any kind. Axis tilting is also out. Aside from perhaps a little bit of global warming or cooling, depending on how the aliens like their climate, anything that would significantly change the Earth would not be in their best interests nor would it be their way to deal with us. They prefer a much more subtle approach and the reason they do is because they know our proclivity for war could be a problem for them. The last thing they want would be us throwing nukes around. We are limited by our technologically primitive nuclear devices that have a tendency to destroy most of the planet and its surrounding space while having no affect on them. This would not be beneficial for us and it's certainly not what the aliens desire. They simply will put us out of the picture entirely with no mess left behind.

Methods that would get rid of humans while still preserving the Earth and all its ecosystems are limited and more than likely have taken time for the aliens to develop. Since we are different from them, they would have to do a bit of research to determine the means they would employ to implement their plans. This, I've been told, is where the infamous alien abductions of humans and cattle mutilations become clearer. They are conducting genetic experiments on humans and animals in order to perfect their diabolical goals. As they plan our demise, they have also decided they need all the Earth pretty much as it is now. That includes all the plant and animal strains that currently inhabit this planet. The last thing they intend to do is upset the exquisite divergent balance of life that thrives and propagates on the archipelago that is Earth. They want every ant, bee, lizard, fish and bird to continue on with the business of bringing forth the essence that makes this planet unique. They intend for every plant to keep on delivering the life-giving oxygen that makes this a livable place. The skies, the wind, the seasons are all to be as they are. The only thing they need to do is eliminate mankind. Right now, the aliens are ending their process of genetic experiments. That is, the abductions and cattle mutilations. I predict these events will drastically decline in the years to come. Not that they won't still take place, but the reported incidences will be far fewer than in the past. This means the aliens are nearly finished with their "lab work" and are about to fully execute their plan in earnest.

They really don't want us dead, physically, that is. No, they are much craftier than that. They don't want Earth to become a contaminated planet, but they do want us conveniently out of the way. The cleanest and easiest technique for accomplishing this goal is what they'll use. I have talked at length with the astronauts about the alien method of ridding themselves of what they consider a pest, but we call humanity. What these unseen forces want is for us to become docile or mentally incapacitated. That of course means they have a use for us. Perhaps they need us as slaves to mine the Earth or as soldiers to help them conquer the rest of the galaxy. Maybe they plan for us to eventually take the place of their robot or android forces. They are no doubt millennium ahead of us in choosing the best methods for rendering a society like ours dormant and totally incapable of defending itself. It is being done by chemically altering genetic strains of our DNA in order to reduce the human function to its basic form and then delivering this potent cocktail through the reproduction system of every human on earth. This "genetic engineering" is accomplished without us having so much as an inkling of what is taking place and we will never know because by the time we could possibly detect such an operation, we will be too far de-evolved to even know or care. This is scary stuff and it is as real as it gets. They don't care one iota about what is right or wrong. Their ethics are nonexistent and ironically, even though they act like intolerant thugs, they could very well be religious and truly believe their God, if they have one, is guiding them to a new world. Maybe they feel they are Gods themselves that have come to punish us for our transgressions against each other and against the Earth. I don't know, and the astronauts didn't know what we are really dealing with here, but I know we are in deep trouble with a formidable foe that seems to have an almost unlimited amount of time to deliver what will be a final blow.

Time to them is probably not what time is to us. They may live much longer lives than we do and a thousand years to them might seem like mere days to us. This would account for their patience in establishing their new dwellings and why their plans for a takeover seem to move so slowly. Time to them may also have other implications. There might be inter-dimensional travel involved that could either speed things up or slow things down for them. One thing is for certain, time is definitely on their side. They've waited, for what has been a long time from our point of view, to devise a scheme that will plunge us into darkness and whatever their schedules are, time for us is almost up. They decided long ago that our Earth is the planet for them and we are like bugs that will eventually be ground into the soil. Their weapon of choice is sublime for its simplicity and all-encompassing delivery capability. It is not a weapon made of steel or iron. It doesn't blow up or make noise. It, in fact, doesn't even kill you or shock you with pain. You won't know when (or if) you've been hit with it and if you do become a victim, you may not ever know it. They've developed the weapon right here under our very noses by abducting and using many unfortunate people as sources for their genetic experiments. Through these experiments, they have gained certain knowledge of our DNA that allows them to splice essential genetic materials with manufactured genes that are, except for their subatomic construction, the same as the genes they replace. These new molecules promote genetic mutations that evolve with time into diseases that affect human brain functions. In time we will virtually re-evolve into a primitive state much like an ape or chimpanzee. Since this method of "taking care of the problem" is in the genes themselves, it will take no more than five to six generations for the mission to be complete. Right now I would estimate between 8 and 10 percent of the human population of Earth has already been infused with the alien manufactured gene and every day the percentage goes up.

# Part Five: The Evidence

Shielded from public scrutiny, astronaut mental problems have been accumulating for quite awhile. Few people know the actual reason why astronaut, Buzz Aldrin suffered a "nervous breakdown" (plus a bout with alcoholism and depression) shortly after returning from his Apollo flight to the moon. Fewer still know the reason behind astronaut Charles Brady's suicide in 1996. NASA reported; he took his own life, citing a "mental breakdown" as the cause. The real reason for these tragedies, unfortunately, goes much deeper than any of us can imagine. Neil Armstrong, the first astronaut to ever step on the surface of the moon, was an outgoing and extraverted individual who loved everything about venturing into space including the publicity. That was before his lunar flight. Afterward, he shunned the public and became a virtual recluse who rarely made any appearances or public comments. Why? In still another case of astronaut mental breakdown, what could possibly cause astronaut, Lisa Nowak, a seemingly stable, successful professional, to suddenly go off the deep end and apparently proceed to carry out the stalking, planned kidnapping and possible murder of her imagined rival? What stresses could have built within this disciplined and pragmatic woman that would cause her to act so irrationally, moving her past the limits of sanity? What put her on a 900 mile, diaper wearing journey from Houston to Orlando to "scare" someone? These stories cause us to wonder - is it the long hours of training …the dangers …the responsibility? - We all have these in our lives. But perhaps Lisa - as well as the other astronauts have something else in their lives …a bit of knowledge about the universe that most of the world only sees as fantasy. Even more curious, in Lisa's case, is the fact that NASA Chief Astronaut, Steve Lindsey (manager of the Astronaut Corps at the time) immediately flew out of Houston in a T-38 training jet to "rescue" her. Now how many Americans could depend on their employer flying a top manager out to over-see their "health and welfare" after committing a felony, almost one thousand miles from home? Why the NASA loyalty? The answer is; Lisa possesses unique knowledge that cannot be compromised. Lisa and most of the other astronauts know of a world changing truth that if brought forward, has the potential to rock the foundations of civilization to its core. I've learned from working with the astronauts of Skylab that even though they are regarded as super heroes by most of the public, they are none-the-less simply people with the same fragile emotions as everyone else. They can, and do, crack under the pressure of knowing the most highly guarded secrets of all time and in these cases that's exactly what happened. These, and other types of mental problems are not peculiar to just astronauts.

There are more people living on the Earth now than ever before in history. This fact would suggest that there has to be more geniuses, more bright minds living than ever before. Yet, where are all the discoverers of new ideas? Where are the visionaries in the sciences, arts, and medicine? Instead we have increasing acts of terrorism and bombings throughout the world. We have the threat of wars and the proliferation of nuclear weapons. We have more people who seem to get satisfaction out of attacking the innocent. We have a rash of school shootings where otherwise clear thinking minds have seemed to snap into an incomprehensible state of evil and the killing of other humans becomes their intire priority and focus. Why?

When one of the world's largest religions teaches devotees that it is proper to kill nonbelievers and if they happen to kill themselves while killing the "infidels" they will earn a very special place in heaven, then humanity is in real trouble. The zealots that pronounce such radical and erroneous interpretations of Islam have proliferated greatly within the last few decades. Why are they suddenly flourishing now? The problem seems to have increased to the point where there are no solutions and, I believe it will only get worse. Wars have not, and will not, solve this atrocity. It is as if men's minds have become clouded and perhaps they have. As little as a half century ago, Islam was a peaceful, well respected religion without any of the fanatical and outrageous dogmas that have now shown up. What has caused the mind of Islam to change relatively overnight? The answer is the same thing that has caused humanity's ever increasing inclination for war as an answer to problems. War has unquestionably increased throughout the world in a relatively short time. For sure, we have always had wars in one form or another, but wars on all scales are now expanding to a degree that is totally out of proportion when compared to past conflicts. It is as though we are evolving backward. We not only find it impossible to solve our problems, but we don't even recognize what our problems are. We have resorted to killing each other, just as it was done thousands of years ago, instead of talking and mutually resolving disputes as you might expect a civilized society would do in the twenty first century. I contend, it is the degradation of mans thinking processes that is dragging us backward and will eventually take us into oblivion.

What could be interpreted as good news is the undeniable fact that on average, human beings are living longer, healthier lives than they have in the past. At first glance this seems to run counter to the allegation that we are degrading and as a species are becoming incapacitated. Living longer and healthier, even living with average higher IQs, does not mean living smarter lives. I continue seeing so-called intelligence moving in a counterproductive way. We are probably living longer in order to have fewer transmission frequencies of genetic material. It takes fewer generations to produce an end result when we live longer. Which leads me to believe the altered DNA strands may degrade with each generation. Their (the alien) goals are being met by our longer lives. As far as having higher IQs than in the past, just turn to the news on your TV (if you can find real news) and try to tell me human intelligence is rising. Film and Television in general have both been on a steep slope toward depravation. TV programming has downgraded sharply into a form of unintelligent voyeurism as, so called, "reality shows" have dominated much of the television landscape. This is no accident. TV executives know their markets and they're giving people what they want. Where are the great dramas and clever comedies that prevailed in the past? Instead we have shows that thrive on sexual exploitation and violence. Where are the imaginations of great writers? Where are the likes of William Faulkner, James Joyce or Tennessee Williams? What has happened to the fantastically brilliant Broadway plays that defined the city of New York so many years ago? Can you name the equivalents of Rodgers and Hammerstein today? Music nowadays, in the form of "rap" has withered into a degenerate expression of negativity that is detrimental to us and our children. Music once was a magnificent art form that moved one's sole. Now, so called "rap artists" shriek out unintelligent howls that amount to nothing more than noise. Strangely, these "artists" know it's just a lot of noise but the public is going for it by buying this junk. It's as though we are hypnotized into accepting mediocrity. Yes, the Golden Age (of everything) has come and gone and we're left with scraps of despotism and decay.

Are we healthier? Compared to the past 100 years, yes - aside from the obesity factor, which is generally an American problem, we are living healthier. However, health has little to do with our brain functions and a whole lot to do with the way medical procedures are patching us up. If you think about it, living longer and healthier means...since we're living longer, we're not dead. So, we are healthier than being dead. Our living longer is exactly what the aliens need to carry out their plans for us and they could be implanting a longer living feature into the human gene pool as part of their scheme. Again, less genetic interfacing between generations means less chance for genetic breakdowns. It all works in favor of those who have plans for us and need the time for those plans to take place

Where are all the great leaps in science that should have taken place by now? It was way back in 1975 when Bill Gates got together with Paul Allen to form Microsoft Corporation. Now they surmise much of their success comes from what is called Moore's Law - the fact that chips (integrated circuits) double in capability every 18 months. They, along with Apple Computer's Steve Jobs and Steve Wozniak, all of that same era, are beholden to an internally known phenomenon that entails, once computer language is established, it then becomes only a matter of time until chip design (using computer functioning for development) evolves into more capable systems. This actually takes much of the human factor out of the development equation.

In reality, digital technology - in essence - improves by itself. Aside from these (automatic) advances in digital computing capabilities, what has been accomplished recently by the human mind?

Where is the cure for cancer that has been coming for the last seventy years? Where are the cures for M.S., diabetes, heart disease, or even the common cold? Instead of finding cures for sicknesses, we seem to have more plagues and viruses popping up all the time. AIDS, Avian Flu, H1N1 (Swine) Flu, Mad Cow Disease, West Nile Virus, Murtha and Ebola are just a few of the names that come to mind that were unheard of just fifty years ago. Alzheimer Disease and Dementia are exponentially increasing to a point where they are almost expected in old age. Autism in children has increased astronomically over the last two decades and now, one in every 150 newborns has the disorder. ...One in every 94 boys. In 1999 the ratio was one in every ten thousand newborns. Autism is a disorder that affects the communication and social abilities of its victims in varying degrees. First identified in 1943, there are now more children diagnosed with autism than with diabetes, cancer and AIDS – combined. Autism is by far the fastest growing developmental disability in the world and there is no known cure.

A report compiled in 2000 by The American Council of Trustees and Alumni, showed that 81 percent of seniors from the nation's top liberal arts and research universities, failed the basics of high school level U.S. history. That's four out of 5 U.S. college students. The council titled the report; "Losing America's Memory: Historical Illiteracy in the 21st Century." Adding to this educational debacle is the finding by The College Board that SAT scores, nationwide, have recently taken their largest drop in 31 years. Look for them to drop further in the years to come.

In another case, a study conducted by an office of the U.S. Department of Education, found about one-third of people living in Washington, D.C. are considered functionally illiterate. That's an astounding fact considering the amount of capital that is expended for education each year. Equally astounding is the finding that one-fifth of the national population is also considered illiterate. These are growing numbers and no one seems to have an answer for this mystifying problem. I think you can guess what the true answer is.

Miscarriages and birth defects continue to be a problem decades after we vowed to wipe them out. Where are the Dr. Jonas Salks and the Madam Curies? More than two thirds of the world still lives in poverty, yet we tend to think we are an advanced society. What have we done to control crime? Child abuse is continually on the rise and violent crimes are on a rampage. The word "terrorism" is now infused in our daily lives and it permeates the news. Where are the great inventions? The computer chip (or, integrated circuit), the laser and structural carbon fiber were all back-engineered in the 1960s from two crashed alien robot craft. You may have heard of the incident. It was near Roswell, New Mexico, (more on this later). We have developed, modified miniaturized, and digitalized many components that have a basis in these alien technologies but as far as real breakthrough inventions go, there's been nothing. Perhaps mapping the human genome was the last great scientific achievement, or the discovery of DNA, but even those were achieved a while back. I suppose once the Large Hadron Collider (the world's largest particle accelerator - located at CERN, near Geneva, Switzerland) gets underway - there will be other discoveries of great scientific value - but, I believe if you look at the whole picture, the age of invention is winding down for humanity and that is not a good sign.

Where are the great visual artists? I'm talking about the likes of Michelangelo, Donatello, Rembrandt, and Leonardo DaVinci. Today, cartoonists are considered the best artists. Where are the designers? We were once promised a fantastic future. Remember the House of Tomorrow? Where is it? Remember the cars we thought we would have in the 21st century? Automobile design is an excellent example of just how far backward we have come in the last several years. On the Engineering side, cars are definitely better. I believe this is a rare exception brought about by computer aided design and the intense competition that has developed as a matter of survival among the car companies. On the artistic side we are stuck in a rut of no imagination.

Cars either resemble boxes or have become so generic you can't tell one from the other. The backward thinking of most car manufacturers has, as a matter of fact lead to what is called "retro designed cars." These are cars that resemble designs of the 1940s, 1950s and 1960s and the really sad thing about these regressed car designs is - the public is buying them like crazy. This alone should tell us something about the public's mental state. Are there no designers left who have at least one shred of imagination and ability to design a truly exceptional new car? My answer is no because their brains, along with ours, are being fried in a way that will cause even the most talented to struggle. We should not be surprised at the title of Lee Iacocca's book..."Where Have all the Leaders Gone?"

Another instance that serves to support the case for a gradual decline in human intellect is the apparent lack of new (non-technical) entrepreneurial skills. Is there a reason there have been no new national fast food franchising successes in the last 40 years? Throughout the mid decades of the 20th century these businesses sprang up like weeds blanketing the country. It has now been over 40 years since this business boom took place and there has been nothing like it since. Where are the businessmen who are capable of starting bootstrap companies such as McDonalds, Wendy's, Jack-in-the-Box and Burger King? Where are the geniuses that created the likes of Domino's Pizza, Taco Bell, Pizza Hut and KFC? Where are the new Sam

Waltons today? Where are the new ideas that should become the business successes of tomorrow? I'll tell you the reason for this unprecedented lull is a complete lack of entrepreneurial talent. In spite of more money being available for loans and venture capital, there has not been a single successful national franchising idea in more than five decades - not a fast food, not a laundry, not a carwash - and there won't be because, other than developing high tech and internet business models (that have been derived by way of computers from back engineering and deciphering the codes of crashed alien robot craft) we, as the human race simply don't have the ability to create anything like we once did.

The elusive "Theory of Everything" is the theory that takes the entire universe, both the ultra small realm of quantum mechanics and the ultra large and expanding realm of intergalactic space and combines them into one grand equation. As the years have rolled by, this "Holy Grail" of physics, that should have been discovered long ago, is growing more elusive in its ability to hide its secrets from us. Scientists today have come up with something they call String Theory. It contends that the entire universe is made up of these tiny vibrating strings. The theory has been modified and contorted into various possibilities without any scientific consensus as to its existence in reality. There are theories of a membrane-based universe that might hold potential possibilities as an answer, but here too there is no consensus or factual mathematical data. Ten and eleven dimensional, multi-universes have been postulated but still no equation. It has been over one hundred years since Albert Einstein conceived his celebrated theories that are the basis of space-time and relativity. It has been over one hundred years since any monumental breakthrough has taken place in theoretical physics. Is there no one the likes of Einstein or Isaac Newton alive today? Why are our professors and physicists not attuned to the universe as they seemed to be one hundred years ago? You'll find the same answer for many questions. Why haven't we gone back to the moon in over 40 years and now that we are planning to return, why is it with essentially the same type of equipment we used back in the 1960s? We're even going to the national archives to dig out the old Apollo drawings to see how they did it. NASA keeps hitting their head against a brick wall trying to figure it out. The problems we once solved easily, now seem overwhelming.

Why are we knowingly changing our climate by burning fossil fuels? Why did we explode nuclear bombs in our one and only magnetosphere and, by the way, why did we contaminate our oceans by carelessly dumping plutonium laden scrap from those same nuclear tests? Radiation isn't the only contamination we're putting into our seas.

A comprehensive study by the U.S. Geological Survey has determined that mercury, one of the world's most deadly toxins, has now contaminated every single fish it has sampled from streams throughout the U.S. The evidence reveals this outbreak is due to coal-fired power plants that release large amounts of mercury from their smoke stacks. These power plants, located throughout the world - but largely in China, are polluting at a rate never seen before. The mercury is constantly raining down onto every single living cell on the Earth. This is an unseen pollution that destroys life by naturally converting into methylmercury that can then work its way into the food chain. Once again man has caused a catastrophe that is beyond his ability to control. There is no escaping this man-made debacle that is slowly killing every single life form on our planet.

In another case, researchers have discovered an area in the Gulf of Mexico covering thousands of square miles where nothing lives. It's called the "dead zone" because everything in this area has been killed by toxic runoff from the Mississippi River as a result of using deadly chemicals for farming. The "dead zone" not only covers this extensive area on the surface, but extends all the way to the bottom of the Gulf. In another region, out in the Pacific - one thousand miles off the coast of California, a floating patch of plastic trash called "The Great Pacific Ocean Garbage Patch" is in a continuous swirling pattern that is the size of Texas. Due to ocean currents, these plastics, along with other man-made pollutants, have been confined in this same vast zone for decades and oceanographic researchers believe the contamination problem will only grow larger in the years to come. I am telling you now, as much as I hate to say it, the human race is getting dumb and dumber by the day and the reason for this decline is the alien interference that is taking place as you are reading this. For want of this planet, they are manipulating and re-engineering our very brains and it is being done so masterfully, you almost have to admire them.

Now if these man-made screw-ups aren't enough to cause you to wonder, here is one other unfortunate situation humankind has created. When the Soviet Union launched their first Sputnik on October 4, 1957, there was truly nothing else, except the moon in orbit around the Earth. Now, here we are, over 50 years later with a virtual junk yard floating above our heads. At the moment, the United States is continually tracking more than 10,000 pieces of space debris, 4 inches across or larger, and admits to "tens of millions" of smaller pieces also whirling through space in orbits around the planet. These pieces of left over space junk, some traveling in excess of 17,000 miles per hour, have the potential to cause severe damage to our vital communications, computer, television and military command and control systems. These satellite threatening and nonfunctioning, useless objects make up 90% of all orbiting hardware. Have you ever wondered why we have been continually launching new communications satellites for the last 35 years at a cost of billions of dollars? The answer lies in the little known fact that defunct objects are tearing through working space satellites, on average, twice a day. It is one more reason that Space Shuttle and International Space Station engineers keep their fingers crossed every day. These pieces of space junk include things like a glove and a spatula dropped by Space Shuttle astronauts, a Russian radio transmitter in a space suit (SuitSat), and incredibly bags and bags of garbage containing human waste from both U.S. and Russian missions. Although the United States and Russia top the list of space contaminators, other contributors include: The European Space Agency, France, India, Japan, several companies and now China has sent a missile into space that has purposely blasted one of their defunct satellites causing an untold exacerbation of the problem. Collisions and explosions are adding to the problem on a daily basis and many pieces are eventually falling back to Earth. There have been several 100 pound (plus) fragments that have survived their dive through the atmosphere and crashed, one hitting a woman in Tulsa, Oklahoma. There are currently no regulations or restrictions governing this escalating danger caused by our own junking up of space. To top it off, here's an article from the wire services that should warm your heart:

> *Cape Canaveral, Fla. Nov. 22, 2006 - Russian cosmonaut Mikhail Tyurin is scheduled to strike a lightweight golf ball today with a special 6-iron from outside the International Space Station in a promotional stunt for a Canadian golf club manufacture.*
> *Tyurin's crewmate will set up a camera to record the stunt for the golf club maker, E21 Golf, to use later. The golf stunt is the first*

> *task for Tyurin and U.S. astronaut Michael*
> *Lopez-Alegria during a six hour spacewalk.*

How much damage can a 17,000 mile per hour, "lightweight" (in space?) golf ball inflect? The stupidity continues ...but It's always nice to know how our tax dollars are being spent.

Politically, things have never been worse. The United States is now one of the most unpopular countries in the world. As a nation, we continually paint ourselves into a corner with no planning or strategy concerning a final outcome. How long will this keep up before we feel the repercussions? Can you remember anytime in your lifetime when Republicans and Democrats have been at each other's throats more? Why is it so? Wouldn't you think cooperation would get more accomplished than the childlike name calling that now prevails? Where are the great leaders? Where are the sages? Where are the likes of Thomas Jefferson, Ben Franklin, or Alexander Hamilton? Where are the Harry Trumans, the Dwight Eisenhowers and the Winston Churchills? Why did we allow the detrimental invasion of 12 to 15 million illegal aliens into the U.S. within the last 30 years? We are like children who have been left home alone. Our values have plummeted into worlds of escapism. TV, I-pods, video games and the cell phone are now our masters. Some sink into the depths of narcotics and alcohol. Others pursue pornography that, thanks to our so-called technology, has become so easy to obtain. Would you believe one out of every 140 Americans are now in prisons? It's true. Downward we spill, into a darkness we will never be able to escape.

Science and technology might have been our strengths in the past, but as far as being an answer to the problems we now face, you can forget about them. It is a fact that within the last few years many of the world's foremost micro biologists have either met untimely deaths or disappeared into thin air. These people could have been leaders in the fight against the alien's program to render us incapable of controlling our minds. The field of micro-biology includes genetics, nano-technology, and quantum physics. All these disciplines involve sciences that are related to what the aliens are doing to us. Could it possibly be, alien intervention caused these unfortunate experts in micro-biology to meet their fate before they discovered what was happening to our planet? If they were around today, could they have created a defense against the plot to genetically lobotomize us? We will never know the answers. What we do know is; by choosing micro biology as their life's work, they paid the ultimate price and I can't help but wonder how many more will succumb as the aliens carry out their plan.

---

# Part Six: Their Weapon of Choice

To put it bluntly, the aliens have chosen a course of action that is as ingenious as it is despicable. After many years of experimentation by what we have called animal and cattle mutilations, they have found the means to render us incapable of defending ourselves. They have abducted human specimens which they've prodded and probed for samples of DNA and other genetic materials in order to gain insight and data that would help with the development of their ultimate weapon. From their invisible perch in space they have sent their robotic (and /or) android surrogates to perform all the necessary preparations ...the dirty work that will enable them to steal the Earth. It's even possible that they had a preliminary test of their weapon sometime in the past. It was a time when human evolution seemed to sputter. Little or no progress was made in any human endeavor. The sciences took backward steps into mysticism and the occult. Plagues flourished and the arts were nonexistent. We called it The Dark Ages. If the alien plan succeeds, a new Dark Age is just a matter of time.

Over all, the evolution of human brain functioning, when compared to other primates such as chimpanzees, is slowing at a surprising rate. These are the latest findings of University of Chicago geneticist, Chung I-Wu. He and his colleagues found genetic brain evolution in human lineage has slowed down considerably when DNA sequences of genes expressed in humans were tested against those of monkeys. The resulting data indicates that while other primates are evolving at an accelerated rate, human brain evolution is actually slowing down. The reason for the difference remains a mystery to science (but now) ...not to you.

The slightest changes in genetic code can make a huge difference in human embryonic development. The brain is uniquely susceptible to any modifications at the cellular level. Many factors could be involved in the developmental dynamics of these minuscule changes. Chemical imbalances or electrical stimulations could be introduced into the nervous system that would allow certain reproductive functions to falter at critical times in our reproductive cycles. Molecules of various harmful elements simply need to be breathed into the lungs to establish lifelong reproductive anomalies. Remember, the aliens don't want us dead as they have a vital use for us. They want us incapacitated mentally but not physically. As I write this, they are attacking our brains by way of altered genetics. Every day that goes by is a day closer to our demise as thinking human beings. I do not know much concerning the technicalities involved in the deployment of their monstrous weapon. That is to say, exactly how they are getting these gene altering viruses into our reproductive systems. In all likelyhood, their delivery system is airborne. The chemical trails left by aircraft have increased during the last several years as we try to defend ourselves with anti-bio agents. It could also be done through our drinking water, or by way of cell phones, or maybe through TVs, computers, or video games. One interesting argument for atmospheric deployment has to do with common honeybees. It has now been discovered that American honeybees are on the verge of becoming extinct, along with many species of pollinating birds and bats. Could there be a connection here? The bees alone, which are needed to pollinate more than 90 commercial crops in the United States, have declined by 30% in the past 20 years. The reason for their consistent decline remains a mystery. Is it a byproduct of the alien's plan to rid the planet of humans? Your guess here is as good as mine. After all, how can we possibly know the technological details of an interstellar space traveling society that may be thousands of years more advanced than us? I do know we have names for the end result their super weapon wreaks upon humans. Those names are Alzheimer's Dementia and Autism – all conditions that ravish the human brain.

Right now, one out of every eight people, age 65 and older, has Alzheimer's disease. That figure climbs to one out of every two (half the population) for persons age 85 and older. In the last five years there has been a 10% increase in Alzheimer's. Between the years 2000 and 2004, deaths attributed to the disease increased by 33%. The Alzheimer's Association calls this a "dementia epidemic". It is a startling, exponential increase that threatens to bring medical and health care costs to the brink. 35 million people, worldwide, have the disease. The Alzheimer's Association predicts the disease will double every 20 years. By the year 2030 it is estimated that 70 million people will have the disease and by 2050 the amount is expected to swell to 140 million people. These are alarming figures that go far beyond what normally would be expected for an aging population. Thus, aging alone cannot account for this rapid increase. Worse yet, a new report by the Alzheimer's Association contains an unprecedented finding: Up to half a million people *under* age 65 either have early-onset Alzheimer's or another form of dementia. This fact has never come out before. Why is this happening? What could possibly be the cause? Publicly, the answer is "no one knows what causes Alzheimer's and there is no cure". I was told these increases would occur way back in 1971 by the astronauts I worked with and now it has all come true. To put it bluntly: An alien force is robbing us of our ability to think and understand and the proof is in the facts.

The Centers for Disease Control and Prevention (CDC) reports: The frequency of Autism has doubled in the last ten years. Right now, 300,000 school-age American children and many adults are struggling to live their daily lives with this suddenly unexpected epidemic. The accelerating rate of this disease has increased ten times in a single generation and although federal funding for research has more than tripled to $100 million in the past decade, no one knows or even has the slightest idea why the increase in cases is so extreme. The mystery has gene scientists working overtime trying to get a grip on possible causes. So far they have found suspicious spots on chromosomes, 2, 5, 7, 11, and 17. What does this mean? For one thing, they admit they are just scratching the surface when it comes to finding an answer. What it also means is there are probably dozens of genes at work causing multiple types of Autism. Clearly, more time, money and research are needed to even begin to understand the causes of the disease, let alone the reasons for its escalating rates. What could possibly cause genes to suddenly go berserk inside the developing brains of children and fetuses, and can the riddle of Autism ever be solved? Progress is being made in a field called epigenetic research. Epigenones are tags that bind to individual genes. These epigenones are somehow altered by environmental conditions experienced by previous generations and influence genes by way of intergenerational inheritance. They are complex in the way they selectively switch gene actions on or off within DNA strands. More research may give us a clue as to how this is done and if there are any artificial modifications that point to alien intervention at this level. Unfortunately, I think the best we can do now is simply help these Autism "victims" by using therapy and training that will eventually bring more normalcy to their lives. Research must continue because as a somewhat civilized society we must pursue that which will advance the human race. I was told the true cause of autism is not to be found on this Earth. I believe the evidence indicates anomalies like autism and certain forms of dementia are the ultimate weapons of an alien race. They developed these weapons and their delivery systems over thousands of years of conquest and hostilities that we cannot begin to imagine. Degradation of the human brain is their "H bomb." They use it generously with a great amount of purpose and without the slightest bit of conscience. Both the young and the old, and everyone in-between are their targets. They prevail because of their boldness. To them, modification of our brains is doing us a favor. They are keeping us alive, and in a way, somewhat happy. Most of our senses will be intact. It's just our brain's processing centers that are being "bombed." When their goals are reached, we will no longer be able to chemically pollute the air and water. The world's factories will be shut. No smoke stacks from steel mills or oil refineries will spill their contaminants skyward. Our need for cars or any other type of transportation will cease. Quite possibly, your great, great grand children will be nothing more than primitive wanderers who understand nothing of the world as we do today – their brains turned into wormwood. I believe this is the end result the aliens have planned for us. Their human problem will be solved by changing our genetic codes and turning us into a race of simple non-threatening beings that will be totally unaware of them as they take over and loot our planet. They are implementing a multi-generational assault that may not end for hundreds of years and the number of lives affected could be staggering. By the mid 21$^{st}$ century it is estimated that close to 10 billion people will populate the

Earth. If their cognitive abilities are continually diminished, it becomes almost incomprehensible to think of what lies in their future beyond the 21$^{st}$ century. Imagine a trillion rational thoughts that will never come to fruition, a trillion new ideas that will never be realized, a trillion pieces art, poetry and literature that will never be known and a trillion dreams that will simply die.

As it turns out, there just isn't much we can do to stop this from occurring. Our military has been firing laser tracking devices at the robot craft for years now in order to determine their flight characteristics and tactics - without much success. The event at Roswell, New Mexico in 1947 was a rare opportunity, but the information gained from that incident still won't let us turn the tide. By the way, it was actually two robot crafts that did indeed crash near Roswell, giving the United States a unique and unprecedented edge for developing "back-engineered" technology. This is what led to many of the sudden technological advancements that took place in the 1950s and 1960s and continue to this day. Not once has anyone ever questioned why every single one of the new technologies happened to be "discovered" and developed in the United States. Since 1947, those robot crafts have been changed significantly by the aliens and are now far too sophisticated for even our latest military resources to stop. There is one theory, however, that the astronauts of Skylab told me, in the long term, could be our salvation.

# Part Seven: Our Shining Moment

How can we possibly defend ourselves against such brutal and cunning aggression? How do you fight an invisible enemy? The answer is: We can't. We are like rats trying to fight a well equipped exterminator. Their plan encompasses a massive program designed to eventually annihilate our brains and bring us under their control, possibly as slaves, or solders, or sustenance. Yes, let's face it. The really big question is: For what possible reasons do they need us alive? In the past, the biggest reason most conquering armies spared their enemies was to enslave them. Putting them to work at labor no one wanted or cared to do. Filthy, dirty demeaning or just plain hard, tiresome, dangerous jobs waited for those who became the conquered. Is this the fate that is mankind's future? I don't think so, and the reason I don't think so is because they, the aliens, already have an abundance of slaves in the form of robots and androids, not to mention the possibility of many other conquered beings slaving away for them. It's not that using us as slaves wouldn't be a possibility, it just doesn't seem plausible to me. Another alternative for exempting us from destruction is to use us as a kind of zombie army - a form of "Cannon Fodder" for their conquests. This to me also seems farfetched for the same reasons as slavery. Moreover, they don't fight their wars with armies; they fight with pure, and not so simple, technology. That leaves, as much as I hate to talk about it, sustenance. That's right; they might need us for food. Will humans be the cattle of the galaxy? This reminds me of the old Twilight Zone episode where aliens came to Earth..."to serve man." Of course, their altruistic blueprint for us turned out to be a cookbook. This again is a possibility, but do you think intergalactic space travelers who have more than likely been in space for multiple generations have a need for red meat? Wouldn't you think they have found a medical means to provide nourishment for whatever their biological systems require? I know your thinking; maybe they grind up their prey into a liquid form and shoot it into their, uh... well whatever they have that will take a whole lot of needle abuse. Again, a possibility but I really think they are beyond our concept of food. Who knows, they may have adapted to receive energy directly from space itself in a form that we know absolutely nothing about called Dark Matter or Dark Energy. This assumption is, in fact, my best guess for their ability to sustain themselves without the primitive and crude habit of literally stuffing meat into themselves. Are we now left without a good reason for them to keep us physically functioning but mentally deprived? Yes. That is except for one outside possibility. This is a conclusion that I have come to only after many hours of discussing the strange realities regarding this issue with the astronauts. I'll admit it's weird so you may not agree with the likelihood of the option I'm about to disclose. Weird or not, I believe it is a promising prospect and it does give us just a hint of how humanity might come out of this - just fine.

I mentioned once before, these entities from somewhere else may not only travel through space, but may also travel inter-dimensionally as well. That means they may be able to travel back and forth through time. Scientists today believe this universe, the one we reside in, may be just one of an infinite number of universes and everything is made up of tiny vibrating strings. This theory is called "String Theory" which I have mentioned before. These tiny strings inhabit ten or even eleven dimensions within this multi-verse. Our senses, in their limited capacity, are only aware of three dimensions which are height, length and depth. Time and perhaps scale can also be considered dimensions, in which case there are five (or six) dimensions that we sense. That leaves five or six dimensions that we have no ability to know anything about. We only know they are real because mathematics tells us they must be there if string theory is genuine. So let's assume our space traveling "friends" have the technology to travel beyond our three dimensional concept of space and can actually travel through multiple dimensions including time. Now they can be from anywhere in the ten or eleven dimensional multi-verse our minds can imagine. On this level our entire Milky Way Galaxy becomes so inconsequentially small and mundane that its significance is practically nil. A question arises; Why would supremely superior beings from somewhere in an astonishingly vast, unimaginably large multi-verse travel through time and space just to pillage our little planet? Could it possibly be - we are that important? I don't think so

...at least not to just any beings. We are, however, very important to one particular life form that most certainly would have an interest in us and what we are doing to our planet. That life form is our future selves. As one astronaut was prone to theorize: "Those nasty bastards could be none other than ourselves from a distant future that, possibly because of some future disastrous event, are going to put us to sleep for a while in order to insure their own reality takes place." I can't help but agree. I believe the aliens that are here and seemingly out to destroy us, are in all probability, our evolved posterity. They are incarnations of ourselves from a far distant future and their very existence hinges on the success they have here and now. If this sounds like science fiction, you're not too far off the mark. But, unlike Si-Fi, this is absolutely real and happening now. You are the first of humankind - that has no special security clearance - to possess this knowledge and what I'm about to tell you will give you at least some warning of the strife ridden years ahead.

All you have read in these pages are true statements that have been conveyed to me by individuals who have had access to numerous confidential and top secret files that exist within our government's military organizations. I have expanded on some accounts and voiced my own interpretations of others. In truth, the basic views of those who voiced these highly classified bits of information have remained as told to me. The truth is indeed stranger than fiction. I have been told we are being subjected to a barrage of genetic manipulations that are affecting the way our brains process information. Aliens who stealthily travel by us, imbedded in asteroids are doing so while sending robotic crafts to Earth that carry out their plans and we will never become aware of the assault. As a possible scenario, brought forth by one of the Skylab astronauts, these "aliens" could very well be our future selves who are in fact here to put us to "sleep" for a period of time because our actions are becoming detrimental to our future, which will definitely affect them. They are temporarily terminating our thinking processes in order to change our behavior to a more altruistic condition. This is not the first time they have altered us by way of genetics. They were responsible for many changes in human evolution. In some cases the changes were for the advancement of mankind. In others, it slowed or reversed human progression. As an example, the emergence of dominating factors inherent in modern humans over those of the Neanderthals might have been their doing and it accelerated human progress. At the opposite end of alien interference is the period of time we refer to as the Dark Ages when human advancement went on a 500-year vacation. The reason they are here at this time in our development is simply because we are at the threshold of destroying the Earth. Our thought processes have not evolved in step with our technology and they are here to put a hold on both, while they fix the problem. They have been constantly monitoring us and have discerned all our irresponsible activities. They've seen all the negative results we have put upon on the Earth. They have given us countless chances to change on our own, but time has now run out. Our imposed "brain surgery" is at hand. How long our "sleep" will last is anybody's guess. If this scenario is true, I suspect it will be quite some time before we once again become technologically accomplished, but that time, although spent in a kind of limbo, will mean we, as humans will survive. Just the fact that our future generations are here in our present and their past, supports the certainty that we made it through this trying time. They are doing what they have to do for their own survival and because we now know they did survive, we might be able to rest a bit easier. These aliens may be our Great, Great, Great, Great, Great, Great Grandchildren. In the end, they will know and appreciate our sacrifice for them. They are the pragmatic ones. They had the means to change their existence by changing their past. They, as it may well turn out, are the ultimate tribute to human ingenuity and intellect. We should be proud of their will to survive and bestow them with honor for winning what must be their greatest battle, the battle against themselves. Their continued existence means we all won. Perhaps their response to a problem they knew must be solved became their salvation and our shining moment. I just wonder if they're (we're) still calling it Manifest Destiny?

Chapter Two

# The irreversible damage caused by the atomic blasting of Earth's Van Allen Radiation Belts.

## Part One: Just Casual Conversations

Much of what follows in this book primarily deals with the question; -Why are the aliens activating their plan now? The answer will soon become self-evident simply by reading the chapter titles. But before I bring you the details, I'd like to explain a little about how information for this book was gathered, and at the same time, bring into focus other topics that came to light during my days as a co-worker and, as it turned out, a confidante to the astronauts of Skylab.

During my 38 years of working in several different facets of aerospace engineering and production, I became employed, and unemployed at various times, for three major aerospace companies. Those three employers were: McDonnell-Douglas, (which is now part of Boeing), Rockwell International, (which is also now part of Boeing), and Hughes Aircraft Company, (which is now part of Raytheon). All three of these companies succumbed to their competitors after I left them, so naturally I like to think the reason they were gobbled up was because I wasn't there toiling for them anymore. This, of course, is not the case and in all honesty they probably missed me as much as you or I would miss a hangnail. That's the way things are at big companies, to them everyone is expendable. Although I retired from Hughes Aircraft, it is my time at McDonnell-Douglas that this book is about. In all, I spent 11 years at McDonnell-Douglas and except for a short stint at the Kennedy Space Center at Cape Canaveral, most of that time was at their facility in Huntington Beach, California. Before Skylab was envisioned, I did work at Douglas Aircraft's facility in Santa Monica, California. There I helped build various missiles and rockets including the Saturn IV and the Saturn IVB rocket stages. It was 1962 when the giant sized Saturn S-IVB, (which was destined to become the third stage of the Saturn V moon rocket) started development in what was then a brand-new facility in Huntington Beach. I transferred there and began working with both electrical and mechanical development teams on the new "Bird." My work followed the S-IVB through development, through production and eventually into test and checkout.

The VCL (Vehicle Checkout Laboratory) included a complex of towers that were built to hold up to four S-IVB space vehicles in vertical positions. A multitude of cables ran from the two south towers into a large room filled with councils, recorders and several of the huge eight bay "Control Data" computers of the day. Production and testing of the various S-IV and S-IVB stages went on in the VCL for the next eight years. Skylab came about almost as an accident. It seems one of the fuel tank sections for an S-IVB ended up with a faulty weld in one of its seams. The discrepancy was discovered after the tank came out of its welding fixture and additional post welding work had already been started. It was therefore deemed irreparable and a new tank had to be fabricated. This left the company with a tank that could not be used. They decided that instead of just scrapping the tank, they would use it as a mock-up, complete with floors and rooms, in order to sell NASA and Marshall Space Flight Center's Wernher von Braun on the idea of using an un-fueled S-IVB as an orbital workshop. It worked and Skylab was born. Once again, as an engineering technician, I was fortunate enough to have worked on the "bird" that was Skylab, from its inception, through its development and into production and check-out. Testing was done in the same VCL facility that had been used for the now completed S-IVB program. In 1970, the two south towers were modified with additional floors and clean rooms. Then, after most production was finished, two Skylabs were hoisted into their vertical homes for what would be just a bit more than two years. Both vehicles were identical with the exception of one being six to eight months ahead of the other in production. The first "bird" was the actual flight article, while the other one was built as a back-up in case something really bad happened to the first one. Excluding a couple of not so minor glitches, the first one performed well in space so the back-up was never used. Today that second vehicle may be seen standing in The Smithsonian's Air and Space museum in Washington, D.C.

At the time it was built, Skylab was NASA's largest and newest program. It was dubbed the most complex engineering system ever put into orbit and to keep the program on schedule, the company and NASA decided to bring many of the engineers and technicians based at The Cape in Florida to Huntington Beach so they would be totally familiar with the vehicle when it was delivered to them for launch. These "Cape Apes"- as we jokingly called them, helped in all aspects of final production and check-out. This program of familiarization worked so well and impressed NASA so much, they decided to do the same with the actual astronauts that would be on board the spacecraft during flight. So, during test procedures and simulations, it was the Skylab astronauts themselves who were flipping switches and reading meters. We, as the technicians and engineers who actually built the vehicle, quickly became friends with our new co-workers. The astronauts were the folks we worked with on a daily basis. We went to lunch at the same places and sat at the same tables. While most lunches were taken at the company cafeteria, occasionally we would all go out to either The Twin Palms Café, or a little beer bar near the plant called Hi Roy's. It was great fun to chat with them over a beer and enjoy their insidious sense of humor.

Anyone who has ever been part of any large space craft check-out effort knows of the holds that can interrupt procedures for minutes, or extend into days. There are waiting periods for inspection to update verifications of instrument calibrations. There are waits for mechanical or electrical changes to be written into programs. There are a multitude of things that cause long waits during around-the-clock systems checks. Many of Skylab's functional tests ran throughout the night and into the next day. Even though there were two twelve-hour shifts working, you could still find yourself awakened by a phone call at two or three o'clock in the morning with a frantic test manager on the line begging you to come into work because they were short handed for a portion of the test procedure that was coming up. Many times, after reluctantly dragging myself out of bed and into the VCL - for me it was only a fifteen minute drive - I would find the procedure in a hold that would last at least until the time of my normal shift start. That was truly frustrating. In order to help all concerned tolerate these unexpected glitches and lulls during check-out, the company, to their credit, moved in a double-wide, fully furnished, mobile home that they situated behind the VCL. The intention was to create a comfortable and, I suppose, calming atmosphere for those of us who were involved in the tests. The double-wide had an intercom that provided everyone in "the trailer" (as we called it) with a live audio feed of the tests that were being conducted at the time. Everyone there also had their own copy of the test procedure, which was large enough to be mistaken for a telephone book. By following the test over the intercom and as it was written in the procedure, you could tell exactly when you would be needed to do your part in the test and the equipment you would need to perform that phase of the test. We used the written procedure as an actor would use a script. At the time of these tests, I had been working at the Huntington Beach facility for about six years. So, unlike the astronauts, I knew exactly where just about every piece of test gear was located along with its calibration status and how I could get my hands on it first, before it was commandeered for some other purpose. I soon became very appreciated by the "space guys." Now don't get me wrong. We all worked hard when there was work to do. There were times when I worked seven days a week, twelve hours a day for several months at a time and the work was definitely not easy. Most of the time I was crawling around scaffolding, forty feet in the air, dressed in my clean room garb - we called them "bunny suits" - while trying to solder wires into a connector, to NASA specs - while standing on my head. At other times, I might be pulling 600 pound cables under the floors of the check-out lab. But, when there was "down time" because of a procedural hold, I, along with other engineers, technicians and astronauts headed for the "trailer" until the wait was over and the test restarted. The double-wide was for our comfort so the amenities, while not extravagant, were very good. We had air conditioning, hot and cold running water, bathroom facilities - naturally, janitorial and maintenance did all the clean up. Every morning we had hot coffee, doughnuts and pastries waiting for us. There was a TV and a conference table that seldom supported anything resembling a conference. Most of the time that conference table was the center of the trailer's activity which just happened to be a card game called, Blackjack. The main reason Blackjack (a.k.a. Twenty-One) was popular with the astronauts was because, on their way out to California from The Cape, they would make a stop at Nellis Air force Base in Nevada. Not only is Nellis bounded on the north by the infamous Area 51, where a few of the space guys had "business," but to the south was the city of Las Vegas where their skills at the game could be uniquely tested. Of course, gambling on company property was strictly forbidden even if it was for only a couple of bucks a hand, but who was going to complain when we were all getting paid pretty good money for just "standing-by." In those days, company management had a tendency to look the other way in order to keep everyone happy and things running smoothly. Even the flow of alcoholic beverages around the aerospace business at that time was not that unusual. I can remember walking past the trash containers adjacent to Mahogany Row at Douglas, Santa Monica just after management gave a contract presentation to the Air Force brass and honestly there were more empty bottles of booze in those bins than I would care to count. I, myself never took a single sip of anything except water or an occasional Coke while at work. I have nothing against drinking - in moderation, it just gives me a headache. I also didn't play cards while in the trailer. That left most of my own down time merely sitting and chatting with other members of the checkout team. Those informal chats included many of the astronauts and over a span of two years, myriads of subjects were discussed. The egos of certain astronauts were more dominant than others and it showed as they discussed their various space voyages and adventures. Sometimes loose talking astronauts tried to outdo each other as they talked about programs they had been assigned to in the past. Spaceflight seemed to be the most seriously discussed subject and this, quite naturally at times, lead to topics like UFOs, aliens, and top secret policies of the government. The booze came out at holidays and on occasions such as meeting a milestone in the test program. Any drinking I saw at the time was always done in moderation and I must say - I never witnessed one astronaut take a drink of liquor in the entire two years I was there. In the trailer, boredom was the enemy and socializing is how we passed the time. The candid conversations with the Skylab astronauts that this book is about were as serious and sober as any you could imagine and the consequence of what they said is immense.

Let me tell you what I learned. First, there were many subjects the space guys said they were dubious of or genuinely knew nothing about. Those included so-called crop circles. They had no more of an explanation for them than anyone. The best guess most of them had was that they were entirely some man-made phenomena...A hoax of the first order. "Scientifically," as they were prone to say, "there is no known explanation." Crop circles in the early 1970s consisted of simpler designs than the sometimes elaborate ones of today. My own thoughts concerning crop circles are that the more complex of these phenomena are of unknown intelligent origin. When confronting a mystery like this, the simplest explanation is usually the closest to the truth (Occam's Razor). If you recall, I mentioned the alien beings were probably like us emotionally. That could include a notion of pride and a need to be recognized. They may even exhibit a sense of humor. These emotions would be far removed from our normal interpretations of how they would become manifest. Keep in mind; a comparison between them and us is not like comparing us to early humans. It's more like comparing us to fleas or bacteria. I can only speculate they have an invisible "beam"- for lack of a better term - that puts these intricate designs in tall stalks of grain, either as a way to tell us they are here and capable of dealing with us, or as a warning by their opposition to be vigilant. Another possibility is simply as a prank, making sport of us as they enjoy our bewilderment. I also think the technology they use to establish crop circles is much like that used to "beam-up" cattle for genetic monitoring (we call it cattle mutilation) and to abduct humans for the same reasons. Thankfully, their treatment of humans is much milder than their treatment of cattle. Secondly, the subject of alien bases or structures on the moon came up, several times. The answer they gave to the obvious question, - Do they exist? - was a kind of irritation at being asked such an inane question. If they discovered structures on the moon, they are certainly keeping it to themselves. Not a hint about this one in the whole two years. Personally, I think there might be something that they saw there but what it is, I have no Idea. To their credit, not even the clowning around or the relaxed atmosphere brought this one out. As far as bases or structures on Mars are concerned, I can't remember if the subject came up. It's definitely not in my notes. I think the time was a bit too early. The photo of the "face on Mars" didn't appear until much later and Mars seemed to be too far away to be of concern. I really don't recall a single discussion about Mars. That doesn't mean there aren't intelligently made structures on Mars, I just believe any discussion of the subject at that time was so unremarkable it was either forgotten by me or it simply didn't come up. We didn't know much about Mars in 1972, and as it's turned out, we still know very little for sure.

Now one subject that I do remember coming up frequently was Earth axis - polar shift. As I understand it, there are two different kinds of polar shifts that can take place. One is where the Earth actually tips on its axis, like going from 24 degrees to say, 36 degrees, or 90 degrees, or 180 degrees. In a case like this we would be literally sunk as tectonic plates would be disrupted and the oceans would spill out of their basins. This scenario is not likely to occur, at least in the foreseeable future, which is not until 25,000 to 250,000 years from now. The astronauts I talked to, really were not too worried about this at all. The other type of polar shift should actually be called magnetic polar shift and this is happening right now, in real time as you are reading this. The magnetic poles of the earth continually drift and at times they can suddenly shift from north to south and vice-versa. These shifts usually occur only at certain spots at the earth's surface and a complete switch of the poles, although rare, can happen either suddenly or over a long period of time. It is the rotation of the molten iron core of the earth that causes these shifts and we have very little information as to what the core is doing. Strata in certain rock formations suggest complete magnetic shifts take place regularly and we could be due for just such an event at any time. The result of such an occurrence could be very significant but not as devastating as axis shift would be. Perhaps the world's navigation systems would temporarily shut down as magnetic north could swing to a location anywhere on the planet. Air transportation would certainly be disrupted as would be ships and anything that relies on a compass for navigation. As bad as these seem, the disruptions would only be of short duration and normality would return soon after scientists discovered the cause. All compasses and computer programs would simply be re-calibrated to the new magnetic north position and life would go on as if nothing happened.

These items will give you an idea of the typical subjects that were discussed with the astronauts of Skylab during my time in the trailer. There were, of course, other matters that also came up; Politics, the economy, jokes (of which I can't remember a single one) and the everyday talk concerning the tests that were being conducted on the space vehicle. I remember a particular discussion that came up concerning Skylab's micro-meteoroid shields. These were curved, ¼ inch thick aluminum panels that spanned the main body of the vehicle. The electrical test and checkout team all seemed to agree they were flimsy examples of poor engineering and when expansion and contraction factors were applied, would fail. They never fit well either in their extended or contracted positions and I could see in my mind's eye, these things ripping off during the launch when high pressure air would get underneath them. When we (the checkout team) questioned design engineering about this potential problem, we were told: "This is the way Huntsville (Marshall Space Flight Center, in Alabama) designed it and NASA is not about to change it." Sure enough, one of the panels tore off during the launch, almost costing the entire mission. These panels also served to deflect the sun's heat, so without the one panel, heat inside the orbiting Skylab climbed to intolerable levels. Luckily, a quick fix - by collaborating engineers and astronauts - consisting of a reflective "umbrella" as well as extra batteries brought up by another Apollo crew to restore power lost when a solar array was totaled by the ripped off meteoroid shield, saved the day.

Of all that was said, during those precocious days, five particular topics are lodged rigidly in my mind and in my notes. The first one is the afore mentioned astronaut's encounter (and near collision) with the asteroid that led to their insightful explanation of alien genetic manipulation of our brains and mankind's eventual mental deterioration. The other four are equally disturbing in their ramifications and also help to explain our meddling with the forces of nature that created the situation we are now faced with. I'll proceed to describe them to you, as told to me in strictest confidence by the astronauts. I do feel a bit uneasy as I bring these matters to light because I did swear I would never mention them to anyone. It took me 40 years to finally decide to write this book. Forty years and your right to know the truth have since overshadowed that pledge I made so many years ago and I now believe I've made the right decision. I'll describe the four other events, one at a time.

---

# Part Two: Exoatmospheric Testing

Did you know Greenland's ice is melting three times faster than scientists thought it was? Satellite measurements suggest Greenland is shedding ice at an exponentially increasing rate. The findings are based on more than three years of observations and echo research data that was released in 2006. Taken with related reports from Antarctica, the study suggests that global warming may be accelerating a rise in the level of the Earth's seas. Greenland's ice sheet holds about 10 percent of the world's glacial ice, nearly 600,000 cubic miles of it, enough to raise sea levels more than 20 feet worldwide if it all melted.

Alarming? You bet it is. A 20-foot rise in sea level would mean most of the world's coastlines would be obliterated. Coastal cities would be wiped out and millions of people would become homeless. World economies would plummet with the entire planet being thrown into chaos. Worse yet, several cubic miles of melted fresh water spilling into the Atlantic would disrupt the global sea patterns of warm water flowing north from the equator (the thermohalien circulation, or the global oceanic conveyer) and the result would bring most of the northern latitudes, particularly Europe into a new ice age. Cities like London and New York could find themselves under tons of ice. This change could take place very rapidly (within decades) and all we can do is watch it happen before our eyes. But wait, it gets worse. The journal *Nature* (September 2006) has reported; Methane, a greenhouse gas 23 times more powerful than carbon dioxide is being released from thawing permafrost and the ocean's floors at a rate five times faster than previously thought. Both methane and carbon dioxide are gases that trap heat in the atmosphere. The result is an accelerating cycle of heat that causes more permafrost to melt, more methane to bubble into the air and more glacial melting.

Global warming (ironically, a precursor to an ice age) is only one of the serious problems we face. The depleted ozone layer has been a huge concern for the past few decades. Every year we see wild weather patterns increasing in their ferocity. Tornados, hurricanes, and monsoonal floods have never been so devastating. The now famous El Nino gets stronger each and every year and from what I've been told by those who were in positions to know, other potentially catastrophic changes that the public knows little about are taking place this very minute. Why, at this time in Earth's history, are we suddenly being bombarded by these global nightmares? The answer could very well shock you.

The chaotic climate changes we're witnessing today are a result of several causes, both man-made and natural. On the man-made side, we have traditionally been told that atmospheric changes are due to the burning of fossil fuels which add tons of air born particulates such as carbon dioxide to our environment. These added particles do affect our climate and we need to be very concerned about the atmospheric contamination they create. In addition, our reliance on live stock for food and milk products has surprisingly added enough methane gas to our environment to change the makeup of the Earth's atmosphere. Human intervention in breeding millions of cattle, sheep and pigs on farms and ranches worldwide for dairy and meat products has indirectly accounted for much of the imbalance we see in global conditions. On the natural side, our climate is affected by the Earth's orbit around the sun. Most of the time, the Earth orbits the Sun in a nearly round orbit keeping roughly the same distance from the Sun. But, every 100,000 years or so this orbit is disrupted and stretched by gravitational forces induced by the planets Jupiter and Saturn. It's called the Milankovitch Cycle and in effect pulls the Earth's orbit into an eccentric elliptical pattern around the Sun. The result is a much closer perihelion and a much further aphelion. This means, at its closest or farthest point to the Sun, a greater or lesser amount of the Sun's energy will hit Earth causing global climate havoc. We are now moving toward this point. Another prevalent natural cause of climate change is volcanic activity. In the past, volcanoes have spewed enough combined carbon dioxide and sulfuric ash into the atmosphere to cause the sun to become blacked out for years at a time. These volcanic events, by themselves have thrown the planet into ice ages lasting thousands of years. There is also the case of methane gas now trapped in the ocean bottoms being released because of increased seismic activity. The estimated amount of this methane, if brought to the surface, could negatively affect our atmosphere for centuries. Basically, mankind and nature have both contributed much to the acceleration of climatic changes but, as you'll now see, not always in the traditional ways science wants you to believe. You will now read exactly why we are headed for disaster when we humans messed with Mother Nature.

Man-made and natural causes sometimes combine to create other catastrophic circumstances for humankind. My Skylab astronaut colleagues told me the one thing they feared most during their moon missions wasn't vehicle malfunctions, they had tri-redundant systems to take care of those, and it wasn't meteoroids pelting their capsule, the chance of that happening was very slim, but what they feared most was something they couldn't even see or feel. That one thing was radiation. A solar phenomenon called Coronal Mass Ejection (CME) could have occurred at any time during their flight. CME is essentially a very large emission of harmful radiation from the Sun. This radiation can penetrate and destroy the individual cells of the body. This is the reason many of the lunar missions took place during lulls in the solar activity cycle. If the astronauts were struck with this type of radiation, their unfortunate deaths would soon follow. Ironically, the Earth is struck by these same harmful particles at irregular intervals. At low levels, the Earth is struck constantly. Then, every decade or so, the Earth is pummeled with high levels of energy from very powerful coronal mass ejections. The reason we, here on Earth, don't receive the same fate that the astronauts might have received - is; we have a "built-in" shield that protects us from space radiation, called the magnetosphere or The Van Allen Radiation Belts. These belts of charged particles cause bad things like harmful solar radiation and gamma radiation from far off neutron stars to be deflected away from the surface of the Earth, thus protecting us from what are essentially the death rays of outer space. As you probably know, radiation from space striking the Earth's magnetic field (The Van Allen Belts) is what causes the polar Auroras (Northern and Southern lights). In a previous chapter, I briefly mentioned the fact that the U.S. government had exploded several nuclear bombs in the Earth's magnetic field (exoatmospheric testing). Those tests, done in the late 1950's and early 1960's carried the code names of "Operation Argus" and "Operation Dominic 1" respectively. It was a time when the cold war was at its peak and the Russians, not wanting to be outdone, also launched their versions of high altitude nuclear tests. We are now reaping the whirlwind caused by sowing the seeds of those tests.

It is believed that the Van Allen Belts are caused by the dynamo action of the Earth's rotating, molten iron core. The belts are essentially a very large magnetic field that is produced much like an electric motor produces current. Electromagnetic radiation, the type that makes up the Van Allen Belts, is a whole lot different from the gamma radiation produced by plutonium laden warheads. Nuclear radiation, as produced through either fission or fusion reactions is deadly to any living organism. Radiation is nature's way of destroying life. It is in essence the opposite of life. With that said, let me tell you what the Skylab astronauts told me concerning exoatmospheric nuclear testing and the ramifications we now face.

It started in 1958 in the South Atlantic with Operation Argus. Argus, in essence consisted of three separate high altitude nuclear blasts that were done to examine the effects of charged particles and isotopes when released into the Earth's magnetic field. 1.7 kiloton warheads carried aloft by Lockheed X-17 three stage missiles, were imploded at 100 miles, 182 miles and 466 miles respectively. Previous to this, the U.S. launched much lower altitude tests by Redstone rockets from the Pacific. These tests were called "Operation Hardtack - Yucca, Teak and Orange." They detonated one 1.7 kiloton and two 3.8 megaton devices at 75,000 feet, 252,000 feet and 141,000 feet respectively. As a side note, It was during Operation Hardtack that approximately 300 U.S. Navy personnel were inadvertently exposed to radiation while on board the destroyer, Mansfield. What happened to these unfortunate individuals is still a mystery. The Argus tests were the first to inject plutonium isotopes into an orbital envelope that completely surrounded the Earth. The 100-mile altitude of the first Argus blast more than doubled the altitude of any nuclear test done previously and keep in mind, this was the lowest (in altitude) of the three detonations. Plutonium has a half life of about 24,000 years. This means the plutonium isotopes released into the magnetic fields of the Earth during Operation Argus will last, in their deadly form, for 24,000 years.

There's more - because Operation Argus was just the beginning of a series of unbelievably destructive blows to our one and only protective magnetic shield.

Induced in 1961 by a Soviet continuation of nuclear testing after a long moratorium, the U.S. also resumed its own program of atomic tests. In 1962 Operation Dominic1 was initiated. As a segment of the test series, there were to be five detonations of nuclear warheads at very high altitudes sent aloft by Thor rockets from Johnston Island in the Pacific. Of these five high altitude tests, dubbed Operation Fishbowl, four were failures and only one was launched successfully. One rocket had to be blown up on its launch pad and the result contaminated the entire north half of the island which to this day is still contaminated and off limits. As a side note, the radioactive debris from this failed test, named Bluegill Prime, was unbelievably bulldozed into the ocean thus contaminating the waters surrounding the island, as well as the island itself. The other failures were all blown up after launching but still contaminated large parts of the Pacific region. The successful high altitude shot was called "Starfish Prime" and it exploded a 1.4 megaton warhead at 250 miles altitude. The blast created a plasma "fireball" that could be seen from Hawaii to Australia. The heat from the bomb was so intense it could be felt by the work crews on Johnston Island. This one test is significant because of the relatively massive force of the blast. When you compare the power of the Starfish Prime detonation at 1.4 megatons with the previous Operation Argus tests at 1.7 kilotons, you see a marked difference in the power of the explosions. As Operation Dominic 1 was taking place, the Soviet Union (Russia) had just completed their own series of high altitude nuclear tests at Kapustin Yar, in Southern Russia near Kazakhstan. These tests consisted of detonating various strength warheads, from 1.2 kilotons to 40 kilotons at 125 to 225 miles in altitude. The Soviets concluded that all four tests in the series were successful (what else would they possibly be?) It wasn't until The U.S. Operation Dominic 1 was winding down that the Russians debarked on another series of tests consisting of high altitude rockets bearing nuclear warheads. These tests, held in late 1962, also from Kapustin Yar, exploded three separate 300 kiloton devices up to 290 kilometers in altitude. In all, there were 21 exoatmospheric nuclear bombs detonated between April, 1958 and November, 1962. Unfortunately, the damage inflicted to the Earth's magnetic field in that Four and one half year period will be felt for centuries to come. Shortly after these tests concluded, The United States and The Soviet Union signed The Limited Test Ban Treaty, bringing an end to the thermonuclear testing madness.

# Part Three: The Results

Before the first atomic bomb was tested, several scientists who were directly involved in the project, tried to stop the test because their calculations showed the possibility of a blast so intense that the subsequent chain reaction might ignite the Earth's atmosphere.

Now I'll give you the results of our nuclear excursions into the Earth's Van Allen Belts. As it turns out, the locations that were involved in the exoatmospheric nuclear tests could not have been worse for the destruction of the magnetosphere. The South Atlantic with Operation Argus, The Pacific with Operation Dominic 1, and Southern Russia with the Soviet launches from Kapustin Yar were like ambushing the radiation belts from three different directions. This Omni-directional scattering of plutonium isotopes covered the region with particles that to this day, continue to infiltrate and dilute the Van Allen belt's ability to protect Earth's living organisms from solar radiation. The high energy electron interferences caused by the multiple nuclear bombs are not only affecting all living things on Earth, but are affecting the Earth itself. The Earth's telluric current that influences the magnetic poles has been compromised and this in turn has been, and will continue to be a cause of global warming. A hastening of the magnetic global shift phenomenon is another result we can expect from the tests, and this is just the beginning of the troubles we've caused.

In the South Atlantic today there is an expansive region that is appropriately called; The South Atlantic Anomaly, or SAA for short. Through measurements taken by way of scientific satellites, it has been determined that in this area, the Earth's radiation belts actually drop closer to the Earth's surface. This phenomenon causes higher radiation dosages to occur in that particular area. It is not surprising that magnetic polar shift has already started in the South Atlantic. Is it just a coincidence that this very same circumstance happens in parts of Northern Russia and The South Pacific? Look for these abnormal magnetic patterns to expand in the near future. Higher dosages of radiation, as in the SAA, can lead to increased cases of cancer and cause genetic mutations in all living organisms. In fact, as reported in the June 1976 Journal of Roentgenology, the Apollo and Skylab astronauts who passed through these nuclear bomb modified belts returned to Earth with measurable amounts of bone mineral and muscle loss. Another element that has largely been ignored by the world's press is the indisputable fact that the hole in the Earth's ozone layer over Antarctica has been getting larger over time. When first discovered in 1985 it was thought that the ozone was being depleted because a higher amount of industrial CFCs (Cloro Fluoro Carbons) were being introduced into the atmosphere. Spray can propellants using nitrogen oxides were also blamed. The result was a worldwide banning of these compounds in 1987. Well, it's been over 20 years since the ban took place and instead of a lessening, or at least a slowing down of the depletion, scientists are seeing an accelerating rate to the depletion. They admit the reason is a mystery. The Antarctic hole in the ozone layer has now become an area comprising 10.6 million square miles. This is an area larger than North America and has been described by NASA as the largest ever observed. I don't think I need to remind anyone that the ozone layer protects us from the sun's cancer causing ultraviolet rays. If those scientists would like an answer to the mystery, I suggest they look at the exoatmospheric nuclear tests that took place in the South Atlantic in 1958.

Now I'm not a nuclear physicist and I certainly don't want to pretend that I know how the characteristics of the radiation belts were changed because of these tests, but it was explained to me at the time and as I loosely remember, after going over my notes, It goes something like this: Although it may seem counterintuitive to think more radiation injected into the belts will weaken them, the facts prove otherwise. Interaction between the charged solar particles trapped in the magnetic fields and radiation from nuclear debris, from the explosions, causes a breakdown in field strength by breaking up electron bonds associated with ionized protons and high energy electrons. Once additional gamma radiation is introduced into the belts, the process continues until all the belts weaken to a point where their mass no longer keeps the charged solar particles locked in the fields. The reduced field strength then releases them and they float off into space. The more dangerous form of plutonium radiation being of greater mass stays within the belts and, along with a natural slow down of the Earth's rotating core, causes the belts to dip lower toward the surface. Many of these deadly plutonium particles will eventually lose their orbital velocities and fall to earth causing worldwide contamination. In addition, the gamma radiation from the blasts gives little or no protection from particles originating from space such as the solar wind, and God help us if the Earth is ever the target of a solar Coronal Mass Ejection or a far off gamma ray burst.

With those details out of the way, let's continue with the results down here on the surface. I mentioned the effect radiation has on all living organisms. Gamma rays are of particular concern because, in addition to their deadly radiation, they also cause an increase in nitrogen oxide (NO2) in the atmosphere. NO2 is a toxic gas that, in reaction to gamma rays, has the potential to choke all living organisms in a global blanket of poisonous smog, while at the same time throwing the Earth into a new ice age. Add to this fray, the shutdown of all electronic components and satellites from the electromagnetic pulse (EMP) generated by such an event and you can see the results would be truly catastrophic. This combination of DNA killing radiation, deadly smog, a frozen Earth and the shutting down of our electronics and communication capabilities have now become ever more likely because of the weakening of the Earth's magnetic fields from our misguided nuclear testing during the cold war period.

Sometimes we overlook the reality that Earth's plant life, is essential to animal life. We could not survive without plants for food, shelter and life giving oxygen. Being naturally more solar reliant than most animal life, plants are even more susceptible to strong doses of radiation. Even in the less extreme scenarios, plants would be the first to go and the evidence of that happening is already here. The Earth's rain forests, which provide a large percentage of our oxygen, are being depleted at an alarming rate. Some of this is due to man's intrusion into former jungle locations for the purposes of farming, ranching and building, but the majority of the defoliation is the result of degradation due to radiation poisoning. In the higher elevations of the United States, scientists are mystified by the multitude of aspen trees that are dying for no apparent reason. These trees are not only dying, but no new trees are sprouting to take their place. The trees and plants are dying and although scientists say they are baffled, the reason is clear. The aspens, rain forests, and all plants and animals including humans are slowly being radiated by the double-barreled effect of the Earth's magnetic belts losing their ability to reflect harmful space radiation and the constant bombardment of cell damaging plutonium particles from the lower altitude fields.

In spite of continual advertising for sun screen products, global cases of melanoma (skin cancer) have increased significantly over the last few decades. In addition, our seas and oceans have been pummeled by contamination in the form of mercuric chloride. The levels of mercury, an element that acts as a radiation magnet, found in fish and other sea foods are rising to a point where the U.S. Food and Drug administration has had to issue warnings against eating seafood products. This follows other warnings concerning milk and dairy product contamination due to cattle grazing on radiation contaminated grasses. As mentioned before, there has been a marked decline of honeybee, bird and bat populations that, in turn, are causing a decrease in crops that require pollination to survive. The reasons for these declines are still baffling scientists. The world's coral reefs, the greatest habitats of sea-life, are dying in never before seen numbers. Healthy coral reefs keep the global archipelago in motion by providing life-giving sustenance to a multitude of living species. Scientists at the University of the Virgin Islands predict as much as 60% of the world's coral could die within a quarter century. When the coral reefs die, the oceans die, and when the oceans die essential oxygen giving algae will die. The journal Science, reports the world will become devoid of all seafood by the year 2048 if current declines in marine species continue. Other scientific studies suggest a marked increase in hydrogen sulfates being emitted from the world's oceans. These oxygen robbing particles could eventually trigger worldwide spices extinctions, not only in the oceans but on land as well.

The results of the high altitude nuclear tests of 50 years ago continue to plague us. Global climate change has been blamed a lot on humans contaminating the atmosphere with such things as aerosols and the burning of fossil fuels. In truth, this is just a small part of the reason. Mankind's real contribution to global warming can be directly attributed to the exoatmospheric nuclear blasts that permeated the Earth's radiation belts during the 1950s and 1960s. As our radiation shields continue to weaken, more and more solar particles penetrate to the earth's surface in the form of dangerous radiation. A by-product of radiation is heat. This added heat, along with more direct sunlight caused by the weakened belts, is proving to be an unimaginable disaster for the planet. One result will be the worst storms in history. Worldwide, storms will increase in numbers and intensity. We already see evidence of this happening. In addition, many areas of the globe will experience droughts that will continue for years on end. Lighting will increase, not only in quantity, but also in magnitude. The strength of these new types of lightning storms will cause a proliferation of forest fires and wildfires to occur throughout the world. As you can see, a tremendous amount of havoc is beginning to result from these nuclear tests. Granted, some of these scenarios will take years to come about, but they are nevertheless all on their way to becoming our reality. Most all of the governments on the face of the Earth have been slow to acknowledge the problem and little, if anything, can be done about it anyway. The United States blames the Russians and (of course) the Russians blame the United States. Both governments know the predicaments these tests have caused, but no one is willing to take any responsibility for them. It boils down to what truly can be called a lose/lose situation with no solution in site.

Before I close this chapter, there's one more thing you might want to ponder. Did you know there was a time in the late 1950s when both the United States and the Soviet Union came close to a decision that would have given the go ahead for testing nuclear bombs - on the moon?

# Chapter Three
# Time for a coffee break.

## The Herb Barton Incident

Before I get to other meaty facts that I learned from the astronauts, I must tell you about the Herb Barton incident. I know I said I wouldn't name any names and I'll stay consistent with that rule by stating; the name Herb Barton is purely fictitious. This account, however, is not. I don't know if Herb is still alive, I kind of doubt it since he was about 50 years old some 45 years ago, but if he is still living, I certainly congratulate him for his long life. Over the years, I learned to like Herb but that's not how it was in the beginning when I first met him. You know how right it feels when someone passes you on the highway going 100 miles per hour, then you see them five miles further up the road pulled over by a cop? Well that's the kind of irony I felt with the Herb Barton incident only instead of the end happening five miles up the road, this true story took thirteen years to unfold.

It started back at the old Douglas, Santa Monica plant back in 1962. I had just started there and had a grand total of about six months in seniority. I was working on several projects, one of which was code named DM-20. This stood for Douglas Missile number 20 or as it was known to the public, Skybolt. In those days most Douglas products started with the word "Sky." Incidentally, "Skylab" was the very last Douglas product that carried the "Sky" prefix. The missile, "Skybolt" was of the air to air variety and hardly made it off the drawing board before it was canceled by the government. Consequently, I was handed a so-called pink slip and found myself laid off along with about 5,000 others. At the time, I had been married for just a year, had just moved into an apartment that I could barely afford and my first daughter was all of two weeks old. On top of that my employee health insurance that had just kicked in would now be terminated. Before being employed at Douglas, I had spent several months job hunting so I knew just how difficult the job market was at that time. Needless to say, being laid off of work at that particular point in my life was the last thing I needed. I did have one thing going for me. Before tossing you completely out the door, the company gave some employees two days to try finding work in other departments within the company. So then, competing with hundreds of other employees, I desperately started looking within the company for work. At that time Douglas did much of their hiring by referrals from managers and supervisors, so I had my supervisor write a very nice letter that brought out all my virtues. OK, so it was only three sentences long. It would still get me in the doors I needed to be in. Well it was tough trying the same tactic as all the others who were being laid off and in truth there was simply no place to go. That was until I found out from a friend who taught soldering and welding classes for the company that there was a shortage of people who could weld wires in tiny modules. He and I both knew it would be a breeze for me to get certified as a module welder. As my two days before leaving were almost up, my friend gave me the extension number of the supervisor in charge of the module welding department. The situation was; In order to become a module welder, you had to go to module welding school and no one could go to module welding school and become certified unless this supervisor of the module welding department approved it. His name was Herb Barton. With my letter of recommendation in hand I made an appointment for an interview right away.

If I remember correctly, Herb was a congenial sort of guy who looked at my letter and asked all the right questions. We seemed to get along well and I thought I had all the right answers. This job is in the bag, I thought. Then Herb asked me, "Do you hold a valid module welding certification?" My answer was, "No, but I could easily be certified if I was sent to module welding school" - as all his other module welders had done. Just a technicality I thought. I then explained that I held a NASA certification for every other type of soldering and welding that was done in the plant. Herb went on to explain that for the jobs that were vacant I needed a valid module welding certification and since I didn't have one, he would have to turn me down for the position. At that point I thought, am I going nuts here? - Am I hearing him right? I asked, in the most polite way I could, "Can't you send me to module welding school? It's only a two-day course. I understand that I can't take the class on my own since your approval is necessary. Is there any way you could approve me for the school? Herb's answer was, "This job requires a valid module welding certification and you're not certified for module welding. Do you have a valid module welding certification?" I think he repeated it at least twice.

To this day I can't think of anything I might have said or done in that interview to cause his negative response. I later learned he hired four or five less qualified individuals, who also didn't hold module welding certifications, but were given the opportunity, by Herb, to go to the classes and receive the certifications. At the time I really felt let down and the reason for the rejection still remains a mystery to me. My suspicions are, in all probability, at the time of my interview he had already picked his political cronies to fill the positions and I was a relative stranger to him.

As my eleventh hour approached, on my final day at work, my supervisor came up to me just as I was leaving and out of the blue said; "How would you like to go on second shift with a slight pay reduction?" I jumped at the opportunity and stayed with the company for ten more years. I never saw Herb Barton again in that entire ten-year period although I did understand he was still working as a supervisor in the module welding department all that time. I went on to work on many different programs from designing and building ground support equipment to doing both mechanical and electrical work on various missiles and space vehicles. I even worked for a while on DC-8 and DC-9 aircraft and had a short stint at Cape Canaveral, Florida. My final project for what was then McDonnell-Douglas was the work I did on Skylab. As it turned out, it was the final job for a lot of people. When Skylab was completed there was nothing to take its place so employees went out the door in droves. The one safety net that saved many Southern California aerospace workers at the time was the Space Shuttle. So as McDonnell-Douglas in Huntington Beach was laying off in droves, Rockwell International in Downey was hiring in droves for work on the Space Shuttle and I became one of their new hires.

I had been working at Rockwell for a couple of weeks when the development job I was assigned to required a special mock-up box to fit the Shuttle's wiring. Much of the unique equipment that was to go on board the Shuttle had not been completed yet so in order to simulate the actual flight equipment and proceed with the development, the company fabricated wooden mock-up boxes that had the exact exterior dimensions of the real equipment. These boxes, which must have numbered in the hundreds, were all painted orange and each had its own part number. The boxes were of various sizes and had dummy connectors located in the exact places the real boxes' connectors would be located. They were an important tool for the development of the Shuttle's wiring harnesses. The mock-up boxes all had to be cataloged and stored in a single location and that location had to have a person there at all times to keep track of what boxes were being used and what boxes were available. As I entered the doorway to the room where these boxes were located, I suddenly stopped in my tracks. Lo and behold...there was Herb Barton standing behind the counter, looking just a little older, but sure enough it was him. My first thought was, I guess even politically well anchored supervisors get laid off if there's no more workers to supervise. Now here was Herb in charge of Shuttle mock-up boxes instead of people. I walked into the front part of the room and up to the counter. I waited until he finished checking out a box for another "customer" then he turned to me and said, "Yes sir, what can I do for you?" I said, "How are you doing, Herb? It's good to see you." Clearly he was puzzled. "Do I know you?" He replied. "Yes" I said. "...From McDonnell-Douglas." "Oh, there's so many of you guys over here now, it's hard to remember you all. What's your name?" I told him my name and that puzzled look appeared on his face again. He clearly didn't remember me or the reason I remembered him so vividly. It had been close to eleven years since our first meeting so I suppose forgetting an insignificant interview wasn't all that unusual. Well I could see he was getting busy so I got my mock-up box and as I was leaving I said, "Nice seeing you again, Herb." He echoed, "...nice seeing you."

Now in case you're wondering why I was even talking to a guy who did his best to kill my income and put me out on the street eleven years earlier, I can't really answer that -except to say, it's not in my nature to hold a grudge. I spent three years at Rockwell and during that time I used many mock-up boxes. Herb and I got along fine and I grew to consider him a friend as well as a co-worker. I never brought up anything about our first meeting at McDonnell-Douglas or mentioned the words; "module welding" to him and I'm sure he didn't recall our little interview in 1962 at all. At the end of those three years at Rockwell, Shuttle development wound down and eventually finished. You know by now what happens when aerospace projects end, and Shuttle development was no exception. Both Herb and I got laid off. Actually, I managed to land a job over at Rockwell's B1 Bomber division and lasted about six more months with the company before President Jimmy Carter canceled the B1 program and I permanently went out the door. Between my time at McDonnell-Douglas and Rockwell, I put in a short period at Hughes Aircraft Company's Electron Dynamics Division. I knew McDonnell-Douglas wasn't hiring, so my first reaction after the Rockwell layoff was to contact my former supervisor at Hughes. Luckily for me they needed a few people to build special test stations that entailed many fabrication skills including, of all things, module welding. They hired me back, right away.

As I remember, one evening after being back at Hughes for about a month or so, my phone rang. It was none other than Herb Barton. It seems he had not found any employment after Rockwell and now his state unemployment insurance was running out.

He had heard through mutual acquaintances that Hughes was hiring and since I was working there, perhaps I could give him the inside scoop as to whom he should contact and maybe put in a good word for him. I know there must have been dead silence for the next five seconds, because I didn't actually know things like this happened in real life. It took 13 years but here we were in almost an exact role reversal. I just about had to put my hand over my lips to keep from saying, "Tell me Herb, do you have a valid module welding certification?" Believe me, over the next several minutes of our conversation it almost popped out of my mouth more than once. It was there, at that point in my life that I decided - no, I'm not going to be another Herb Barton. I'm not going to stoop to the same actions he displayed to me those many years in the past. I did realize that the moment was probably a once in a lifetime, absolute pinnacle of irony - but irony or not, I wasn't going to succumb to the temptation. I decided there and then to take the high road and do the right thing. I gave Herb my supervisor's extension number and told him I would help pave the way for him with a positive recommendation. Herb thanked me, we said goodbye and hung up.

The next morning, I told my supervisor (who I knew still had a couple of positions to fill) about Herb's call to me the previous night. I did my best to give Herb a positive recommendation and let my supervisor know of his capabilities. I mentioned that he should expect a call from Herb that morning. My supervisor was pleased as it was getting more essential that these vacant positions were filled. Later that morning, my supervisor approached me to let me know he had talked to Herb over the phone and had scheduled an interview with him that afternoon. I thought to myself - things look pretty good for Herb. It was almost the end of the day when I saw my supervisor again and inquired about how Herb's interview went. He told me Herb came in and they talked for about 45 minutes. I said, "yes...and?" My supervisor then replied, "I just don't think he would have been a good fit for the job." Then he turned and left. I couldn't help but feel just a little gratified. In spite of my doing what was the right thing, in the end poetic justice prevailed.

I never saw or spoke to Herb again after that and I sometimes wonder if during his interview with my supervisor, he didn't get just a twinge of deja vu and finally remember who I was, and our little interview 13 years earlier. The reason I really think that's what happened is because, after my supervisor informed me of his decision concerning Herb, then turned and left, I swear he said in a soft voice; "...and besides, he didn't have a valid module welding certification."

Several months later I heard Herb got his old job back at McDonnell-Douglas and was doing just fine.

Chapter Four

# The nuclear bombs that will soon fall to Earth because of deteriorating orbits.

## Part One: Nuclear Weapons in Space

OK, now back to more fantastic things I learned while working with the astronauts. The conversations I had with them during the check-out phase of Skylab covered a multitude of subjects. The events I'm about to delve into are some of the most compelling. Please keep in mind, what was told to me was only done because I agreed to complete secrecy. These accounts were given by astronauts and engineers who actually worked on the projects discussed in this book and the subjects were presented to me as absolute truth. True or not, all I can do is present to you what my memory and notes recount and invite you to be the judge.

With the Soviet launch of Sputnik in 1957, a new era began for potential weapons delivery systems. The military applications for space hardware were apparent even before the first manmade satellite went into orbit. When the USSR became the first nation to send a satellite orbiting the Earth, the United States' rationale for orbital military use of space became a major priority. It was a time when the Pentagon bristled with plans and scenarios concerning space based military hardware and many of those plans dealt with the pros and cons of placing a nuclear bomb in Earth orbit. At the onset, the advantages of a device like this in Earth orbit seem obvious. No rocket launch would be detected since the device could be launched months or years prior to its use - disguised as a reconnaissance satellite. No expensive hardened silos would have to be built to accommodate and safeguard the weapon. A network of devices in polar orbits could blanket and reach any point on Earth within minutes. The entire program could easily be carried out within other overt or covert military projects and become implemented in complete secrecy. At the time these Pentagon discussions were taking place, the nuclear non-proliferation treaty between the Soviet Union and the United States regarding nuclear weapons in space was well off in the future, but regardless, during those days treaties were ignored by both sides and both sides in the cold war had the intelligence capabilities to know what the other side was up to. I found this out on my own when I attended a top secret briefing at one particular aerospace company I worked for and was told the reason for the super-secrecy was; "...not because we're trying to keep it from the Russians, they already know all about it. We're trying to keep it from the American people. We don't want them to know we're breaking a treaty." The advantages of having nuclear warheads in orbit at that time in our history bode well for those who wanted them there. The disadvantages, while not as obvious were also substantial. For one, many warheads would be needed to cover the necessary areas which would add to the program's cost. A system for maintaining the devices had to be devised which would also be expensive as well as complex. The devices would be vulnerable to enemy attack. Their electronics systems could be compromised by electro magnet pulse (EMP) resulting from an enemy nuclear blast. Clearly the concept had some inherently large issues facing it especially since any future nuclear nonproliferation treaty between the U.S. and the Soviets seemed likely to either outright block or place the program into an ultimately suspicious secret environment. The Pentagon batted all these "what ifs" back and forth, but at the time, was reluctant to phase in such a controversial and potentially provocative program. The real convincing came by way of the Soviet Union in 1963 when, as usual for that time, they became the first nation to secretly place a nuclear warhead into orbit around the Earth. This is all it took for the U.S. to go ahead with their own program to orbit a bomb. The initial idea was to put a single device into a polar orbit. The Air Force would provide the launch vehicle and the launch logistics. Vandenberg Air Force Base would be the site for the launch. The military, however, soon got mired down in an inconceivable argument about who would be responsible for the many details the program would entail and while this bickering was taking place, the USSR launched their second warhead into orbit. At first the Air Force decided they would maintain their weapon with a two-man crew launched aboard a spacecraft they called the Dyna-soar, an appropriate name as this part of the project was soon canceled in favor of another Air Force program that I brought up in the beginning of this book, designated MOL, for Manned Orbital Laboratory. The project consisted of a modified Gemini B space capsule mated to a cylindrical work station and launched by a Titan III rocket. This system turned out to be the one the Air Force would go with. The MOL was sold to the public as an orbital space laboratory where the U.S. could test and develop the military applications of space. The MOL was sold to the Soviets (and the Pentagon) as a manned orbital reconnaissance camera capable of taking high resolution photos of any global position. At the time we believed the Russians were photographing our military installations from cameras aboard their manned spacecraft and we wanted them to know we were about to do the same to them. Both these applications for MOL were valid. The third application, which few knew about, was to control and maintain orbital nuclear arsenals. The MOL, as it began development, was known by various code names including KH-10, and Dorian but as it turned out its life was short. The MOL was canceled in 1969 mainly because other unmanned satellite surveillance programs such as "Corona" became a more viable and less expensive way to do the same job of reconnaissance. For the U.S., the idea of orbital nukes became a moot point as fractional orbital bombs (FOBs) became the favored delivery system. Since FOBs only orbit one time, directly after launching, maintaining orbital nuclear devices was suddenly not necessary. Incidentally, several astronauts who were training for the MOL program, when it was canceled, moved directly into training for Skylab and subsequently became a part of the astronaut group that was the "source" for this book.

There are two ironies we are left with because of these 1960's nuclear shenanigans. The least important but never-the-less very expensive one is; -while we (The U.S.) were building our huge anti-missile defense system that included both long range and short range missiles plus the most advanced radar installations in the world in states like North Dakota and Montana to stop Russian launched ICBMs from coming over the North Pole, (the shortest and most logical route from the Soviet Union) they had their weapons already launched and in polar orbits that could attack us from the south where we had absolutely no defenses or radar systems to warn us. At a time when we thought we were relatively safe and had spent billions on defending the U.S. from missiles launched inside the USSR, we were like sitting ducks. The Russians truly had their nukes pointed at the back of our heads while we looked out the front door. The other irony is even more compelling.

# Part Two: What Goes Up, Must Come Down

On July 11, 1979, Skylab plummeted back to Earth in a fireball. The vehicle broke up over the Indian Ocean strewing large chunks of rubble as it descended. Weeks before the demise of Skylab, there were debates by space engineers and scientists as to where and when the break up would occur. The general time of de-orbit only pinned the event down to a few days and the eventual location of its landing was anybody's guess. There were actually contests put on by radio and television programs that offered prizes to the first person to bring in a verifiable piece of Skylab. All over the world people looked up to see if they could get a glimpse of the space lab as it was tumbling to Earth. The anticipation and anxiety faced by those who thought the huge piece of space hardware was about to crash into their homes was alarming. Finally the news broke: Skylab's demise took place somewhere above the Indian Ocean. Now everyone could breathe easier. A few days later, reports started coming in that described a particularly large part of Skylab half buried in Australia's Outback. It turned out, the reports were true and it was only by sheer luck that Skylab missed any populated areas. The death toll, even in a desolate part of Australia, was one cow.

The results of any large piece of space junk falling on our towns or cities would be devastating. Just think of the possible ramifications if that space junk happened to be highly contaminated with radiation. The fiery decent of Skylab points out a fact that we have faced ever since the space age was borne. Unless properly maintained, the orbit of any man-made satellite or space hardware will, in time, decay. One of the prevailing reasons for orbital decay is atmospheric expansion. That is to say, radiation activity on the Sun will cause our atmosphere to expand to a point where it interferes with orbital trajectories. When you combine this with Earth's persistent gravity, you have the recipe for a descending orbit. Modern communications satellites get around this problem by their higher orbits that match the rotation of the Earth. Hardware in lower orbits must rely on continual boosts from small, onboard rockets and/or other external tugs to stabilize their orbits. The Soviet orbital nuclear warheads have long ago run out of fuel for their small orbital sustaining rockets and no manned trip to fix them will ever take place. It is estimated that the former Soviet Union launched two orbital nuclear devices in the early 1960s. I suppose, to them, their actions at the time seemed to be an excellent deterrent to the nuclear build up of weapons the U.S. was establishing. Whatever their reasoning was, the results of their imprudence are now becoming manifest in what could well be a catastrophe. The orbits of these former Soviet warheads have been deteriorating for the last couple of decades and are now approaching a point where they are a serious threat to the world's natural environment and human population. Just as Skylab's final orbit was unpredictable because of atmospheric and gravitational variances, so it is with these large orbiting bodies. In reality, they should have come down years ago, but the Earth's upper atmosphere has cooled an estimated 18 degrees in the last 30 years. This cooling, caused by the same greenhouse gases that warm the Earth's surface, results in the upper atmosphere contracting and becoming less dense. Cooler and less dense air that is contracting, means less drag on orbiting satellites - in turn, causing them to stay up, sometimes decades longer than anticipated. This is the case with the soviet nuclear devices. It is virtually impossible for space scientists to calculate the precise time or date they will crash to Earth. The location of their impact point is also not definable. What is known is the de-orbit event should occur within the next several years and one warhead could follow the other by a matter of just a few hours - or possibly days - or even weeks. Unlike Skylab which was placed in a latitudinal or equatorial orbit around the Earth, The Soviet weapons are in a longitudinal or polar orbit. What this means is, that since more land areas are covered in a polar orbit, it's far more likely the orbiting bombs will fall on land than it was for Skylab. The really big problem here is, there will be absolutely no warning before the weapons fall. They could hit anywhere on Earth, day or night and there's nothing we can do to stop them. Any intervention, such as blowing them up while still in orbit, by any nation would jeopardize the entire world by spreading deadly radiation globally. As we learned from Operations Argus and Dominic 1, this is not a good idea. Sending any type of manned mission to the orbiting warheads could easily become a suicide mission because it is believed the devices are leaking heavy amounts of radiation from faulty casings. Through highly secret discussions between the U.S. and Russia it was decided long ago that it is best to let these nuclear relics simply fall to Earth on their own when the time comes. The consensus is; the weapons don't possess a possibility of a nuclear blast, even on impact. Although the conditions for detonation have been deemed no longer viable, this does not mean the weapons aren't deadly if they happen to hit even a semi-populated region. The radiation alone from just one of the damaged warheads could easily be a disaster for anyone within a radius of one to two miles of the impact site. On the positive side of all this is still the large possibility that the falling weapons will splash into the sea where the radiation would dissipate more rapidly. We are left with another debacle by world governing bodies that never seem to look into the long term effects of their actions. The consequences of the cold war will be with us for some time to come and these orbital nuclear warheads are only part of the extraordinary overall picture that has been told to me by my astronaut friends 40 years ago. I, in fact, had to check my notes twice when I came across the next part of this account concerning the falling Soviet nuclear weapons. The scenario fits in so closely with what's happening in the world today, it's uncanny. Again, I invite you to draw your own conclusions.

---

# Part Three: Placing the Blame

You must remember, these accounts are of highly sensitive and classified events, some rated above top secret. With that said, you can imagine the dilemma the Soviets and the U.S. had, and still have when the time comes, in trying to explain to the world's populace the reason a couple of radioactive warheads just dropped from out of nowhere and into some farmer's hog pen. There's not supposed to be any warheads in orbit. We and the Soviets signed treaties that strictly forbid any nuclear proliferation in space. Whose warheads are they? Can't you just hear the media's reaction to this fiasco? Well it's not going to happen and here's why.

If the nukes fall into the ocean or any large body of water, there's no problem to explaining them as meteorites. This happens quite regularly in the skies on any given night. If they hit land, then the cover-ups will begin. Right now the United States and the Russians have trained teams assembled throughout the world just waiting to ascend on the warheads with a barrage of heavy shielding and radiation isolating equipment. These teams are well trained and have practiced being dispatched to a multitude of locations numerous times. If this is what occurs, the military's public relations department (yes, they actually have one) will hustle into action in order to convince the world that an alien craft has landed and for reasons of security, they had it whisked away, draped in a cumbersome blanket of shielding, to an undisclosed location. If this sounds familiar, it ought to.

The Kecksburg, Pennsylvania incident is a well known (alleged) UFO landing story that took place in December of 1965. To make a long story short, an object was seen streaking across the sky and crashing to Earth in the woods just outside of Kecksburg. The military, with Geiger Counters in hand, came on the scene almost immediately and kept all onlookers away from the site. Later, a flat bed truck was seen leaving the area with a large shrouded object on board. The cover story the military gave for the incident was, "a meteorite fell to Earth without causing any apparent damage." They of course hoped a UFO connection would be taken up by the public to cover the hauling away of the object and that's exactly what happened. Now, I'll reveal to you the true, top secret account of the Kecksburg incident as it was told to me. This account has never been told to the public before and as far as I know is still classified information. First of all, what do you think the public outcry would be back in 1965 if it became known that the Russians had implemented an orbital nuclear system aimed at the U.S.? We didn't want this to become known and the Soviets certainly didn't want anyone to know. If you haven't guessed by now, that was no UFO that landed near Kecksburg that night. There has been some speculation that the object was a Soviet satellite, perhaps one of their Soyuz capsules that might have experienced difficulties. This account is closer to the truth because what was found in the forest that cold winter night was indeed of Soviet origin.

Fractional Orbital Bombardment, FOB for short, is what the program was called. The Soviet FOB program consisted of missiles zeroing in on their U.S. targets from space after completing one orbit. Naturally, the delivery system needed to be tested in order to become a verifiably dependable way to destroy U.S. cities. The Russians, in their exuberance to again be first at something, launched into orbit an inert test model of one of their warheads to see if the device would perform as they intended. The object was an actual nuclear warhead that had been modified to carry the same weight as the real one. The test was supposed to place the device into Northern Canada near the Arctic Circle. We could only guess that a malfunction in the guidance system is the reason the USSR nuked Kecksburg way back in the winter of 1965. Luckily for everyone on the planet the warhead was only a mock-up. The U.S. has kept the incident under wraps all these years and because they always want to keep their adversaries guessing, so it will probably remain for decades to come.

Now back to the nuclear situation that persists today. The U.S. stands much like it did back in 1965 as far as not wanting the public to know about these, soon to fall, radiation dangers. We can argue back and forth about the wisdom of this approach but, non-the less this is the reality that we find ourselves in. I've covered the warheads dropping into the sea scenario and the military getting to the scene first scenario. Now I'll discuss the one circumstance, both governments (U.S. and Russia) would prefer not to happen at all. That circumstance is, what to do if the trained teams cannot get to the devices first and the crashed warheads cause widespread radiation poisoning and damage among civilian populations. Needless to say this is the worst case scenario and you might even think that if this happens the responsible governments should own up to their desecrations and take responsibility for their past actions. This, I can assure you won't happen because they have a cunning plan in place that is as incredibly clever as it is shocking. The plan has been taking shape for the last 40 years and involves the cooperation of both governments.

As a result of the Cuban missile crises in 1962 and a number of other "close call" incidences of nuclear craziness, both the U.S. and the Soviets began to see the folly of continued nuclear threats. This became the beginning of an extraordinary era of peaceful negotiations between the United States and the Soviet Union. Most U.S. citizens are under the impression that our blockade of Cuba and our threats of nuclear war is what caused the Soviets to pull their missiles out of Cuba. Look in any newspaper of the time and you'll see that story after story basically says, the U.S. threatened and the Soviets blinked. True, the Soviets agreed to take their missiles out of Cuba but that came as a result of negotiations. The reason the Soviets put their missiles in Cuba in the first place was a direct response to the U.S. placement of missiles in Turkey under the guise of NATO. The negotiations were simple. They take their missiles out of Cuba, and we take our missiles out of Turkey. It was what the Soviets were aiming for all along. What we got out of the deal was a highway to future long term talks with our cold war enemy. Several of those future talks involved the plan for explaining how Russian nuclear devices suddenly might appear in places worldwide. Eventually these secret talks with the soviets led to a standard protocol that addressed various subjects including terrorism.

Terrorism has been a threat since the end of World War II. In the 1950s there were sparse cases involving terrorists. Airline hijacking was the biggest threat at that time and those seemed to be carried out by individuals or small groups of deranged thinkers who were obsessed with flying to Cuba or Algeria for political reasons. The outcomes were more of a nuisance than the horrific suicidal situations we've come to expect today. As the terrorists persisted and their misgivings started to include groups with religious ideologies, a plan was hatched between the United States and the Soviet Union. Both countries decided to take advantage of the rise in terrorist activities. At the heart of the plan was the question of, what would happen if a terrorist group got their hands on a nuclear bomb? The consensus of resulting meetings held to discuss just such a scenario was that eventually, given enough time, it was likely to happen. With this in mind, the two governments came up with the answer to the orbital bomb dilemma. First of all it was a U.S. problem because the United States didn't want the public, or the press, to know it became a sitting duck to a possible Soviet nuclear attack at the height of the cold war. Secondly, it was a Russian problem because they lunched these weapons in violation of non-proliferation policies negotiated by entities of which the U.N., NATO and Warsaw Pact countries were all a part of at the time. Both governments needed a scapegoat and it fell into their laps in the form of terrorism. How would we explain the bombs to the rest of the world if the troops couldn't get to them in time and they caused a lot of radiation damage? That cover story is now in place. We will, with Russia's confirmation, simply blame the incidents on terrorists. The stage is set as we are now conditioned to expect a possible radiation or "dirty bomb" to be slipped our way at any time. The perfect cover story has been fabricated. It covers the original origins of the devices. It kindles worldwide sympathy for the fight against terrorism and it gives western forces even more clout to move into the oil rich Mideast in order to oust the Islamic extremists. The event itself may not happen for several years or it could take place tomorrow. There's simply no telling how close these threats are to us and the Earth's environment. What this fiasco does point out is our failure to manage this planet in a responsible way. It is no wonder alien entities need to get us and our disruptive vandalizing of this virtual paradise called Earth out of the way. They have decided, long ago, to take the planet away from us for an unknown period of time because of our lifestyles that show no respect for what we have here. Quietly and ever so slowly we are falling into a bleak darkness that has been induced by the alien plan for our dissolution. The case of the orbiting nuclear bombs, in part, tells us why.

Chapter Five
# The secret U.S. practice of releasing bio, germ and radiation contaminants on U.S. cities.

## Part One: The Rationale for Testing

There are many mysteries in medical science that are difficult to solve. We must experiment and test, then wait for results, then experiment and test some more. This is done to insure the end result will be the best and most promising solution for whatever problem needs to be remedied. Sometimes the possibility of saving a multitude of lives by sacrificing a few can simply seem like part of the cost for advancing science. However, testing on unknowing innocent citizens, without regard for the individual's human cost is not an acceptable means of advancing medical knowledge - or so you would think.

In the early part of the 20 century, America was a different sort of country. It was a time when P.T. Barnum brought the circus sideshow to most American cities and people were mesmerized by what was known as the freak show. Curiosity is what drove the populous to these gaudy expositions that in today's world would spawn outrage and ridicule. But in those days it was the unfortunate subjects of the freak shows who were the ridiculed. Curiosity, as well as indifference is what brought America to Barnum's Freak shows. Except for a few, the people just didn't care much about the "other guy" …even if that guy was debilitated physically or mentally. It was a time when those with mental illness were stuffed into institutions to slowly die degrading deaths. That's just the way it was because most people then were not educated or taught to respect those less fortunate. It was in this environment that the U.S. government began medical experiments on mental patients, prisoners and even children. This was the beginning of what was to eventually become full blown germ, radiological and biological testing on innocent U.S. citizens without their permission or knowledge.

Plausible deniability: What a person doesn't know cannot be discussed by him or her as fact. Thus, it is legitimately deniable. It's why the U.S. president hasn't been told the truth concerning the alien threat to humanity as well as facts behind many other secret programs. The testing of weapons of mass destruction on our own citizens is one of these cases. Over the last few decades, plausible deniability, as a defense, has been evoked for senators and congressmen, cabinet members and lower management government officials alike. What these folks in government have been told are the relatively mundane and expected accounts that deal with development and testing of new military hardware, troop deployments and overall intelligence gathering programs ...all, of course, on a "need to know" basis. They really don't know a thing about what is told in this book. There is a gaping disconnect between our political government and the entities that are often referred to as the black government or the covert government. With this in mind, I'll proceed to tell you of some extraordinary secret programs that have taken place in the past. They are programs that I am simply relaying to you as I scan my notes taken more than 40 years ago. What you are about to read will likely shock you, but please remember, -America was different in those times and our leaders were (and still are) the victims of plausible deniability.

---

# Part Two: Pure Insanity

First, a Question: How would you test a bio or germ agent to verify its potency and potential? Well, 40 years ago this exact question came up while chatting with a couple of astronauts and several test engineers during one of Skylab's lengthy "hold" periods. As I look at my notes from that conversation, I find myself shaking my head in disbelief concerning the decisions our military made in regard to the answer that question raises. What we, the citizens, don't know won't hurt us might have been their credo, but in retrospect the exact opposite was true. Before I tell you the military's response to that question, let me explain that most of what is known about the outcome of any bio or germ tests on U.S. soil has been covered up or in many cases, minimized by the agencies or branches of the military that conducted the tests. Some government sources even deny testing ever took place. This is plausible deniability in action. How did the military answer that question about how to test their bio and germ weapons? They unconscionably and unbelievably decided to release the toxins on our own American and Canadian cities and keep the tests top secret. The real, irreversible and absolutely horrifying consequences of those actions are what I will now reveal to you.

Between the years 1949 and 1969 the U.S. government spread numerous biological and germ agents throughout many populated areas of North

America just to see how many (innocent) people would become infected. During these so-called tests, hundreds of American's most susceptible, such as the elderly and newborn infants had mortality rates that skyrocketed beyond what was normal at the time and no one could figure out why. I might suggest a very good candidate for this anomaly as there were more than 200 separate releases of active pathogens during this time period and all of them were done without the public or the U.S. congress being aware of them. Many of the diseases introduced during these open air tests still linger in regions of the U.S. today and are still infecting people as you read this. North America wasn't the only part of the world affected by these experiments. Both the United States and the former Soviet Union conducted secret bio and germ testing in Asia, the Caribbean and Central and South America. The territories covered by the tests were expansive and in fact some of these contaminants found their way around the world.

I suppose you could say this all started in the late 1700s and when both British and U.S. forces distributed blankets to Native Americans as a gesture of good will. What the Indians didn't know was just how diabolical their new neighbors could be. The blankets, these generous gifts to show our concern, were purposely infected with cholera and smallpox. The Indians had absolutely no immunity to these diseases and this one act alone set in motion the declines in Native American populations that are still being measured today. Keep in mind, my knowledge of this subject is somewhat limited and is mostly taken from the notes I hurriedly took after my conversations with the astronauts. As with all accounts this book contains, I have corroborated this knowledge through research and my own understanding and personal feelings concerning the subjects discussed. It is a combination of these components that brings me to the revelations I am about to disclose. As I've said before, you are invited to draw your own conclusions about what this book contains. I would be the first to admit, some of what is covered here sounds unbelievable but it is what those conversations entailed and I look at myself simply as a reporter who is relaying the words to you. With that said, I'll mention that one astronaut who I spoke with during those days on Skylab was a West Point graduate who knew an extraordinary amount of military history connected with the bio and germ experiments. The notes I took of his accounts greatly influenced this chapter of the book. In order to get a sense for what was done in the name of medical research in the early part of the 20th century, I'll site two cases of unconscionable circumstances.

First, in 1911 the Rockefeller Institute for Medical Research published a report on their experiment to develop a skin test for detecting syphilis. The experiment involved injecting what was thought to be an inactive syphilis preparation into the skin of 146 hospital patients and normal children. Needless to say, the experiment went awry and two years later there were numerous lawsuits by the parents who rightfully alleged their children were infected with the disease. The second case, in 1913, involved 15 children at the St. Vincent's Children's Home in Philadelphia. Medical experimenters knowingly infected the children with tuberculin, causing permanent blindness. The Pennsylvania House of Representatives hears and records the incident but the experimenters are immune from punishment. These two cases set the stage for what follows.

Mustard gas, (or yperite) a blistering agent, was first used as a weapon by the German army in 1917. Dispersed as an aerosol with other chemicals, it was quite effective at slowing down an advancing enemy. The gas not only blistered the skin, but caused painful damage to the pulmonary system if inhaled. After an initial attack, lingering gas would remain in the environment for days and contaminate large areas of land causing "no entry zones" for enemy armies. At least 10 different countries have used mustard gas as a weapon since World War I and, not surprisingly, these instances only took place when it was certain the opposing forces could not retaliate. During World War II, both allied armies and axis powers refrained from using mustard gas (or any toxin) as they knew retaliation in kind would soon follow. This mutual "respect" for one's enemy became the precursor to the MAD (mutually assured destruction) policy that was later implemented between the United States and the Soviets in the nuclear era. After WWII, Germany's cache of mustard gas was dumped, by the allies, into the

Baltic Sea. Now when mustard gas is exposed to water it forms a deadly tar-like gel. This substance stays lethal to humans as well as sea life for several years and the prolonged effects of this disposal have yet to be fully understood. There is no way to tell exactly how long the sealed containers that were dumped will retain their integrity.

In 1972, the year Skylab was being built, the United States Congress banned the dumping of U.S. stockpiled chemical weapons into the oceans. By this time, however, the army had already dumped more than 64 million pounds of nerve and mustard agents into the coastal ocean waters of the United States, knowing the contamination problems this would cause. If you think this is unbelievable, read on. According to the U.S. Army Chemical Materials Agency, there were at least 26 chemical weapons dumpsites located off U.S. coasts. These sites include Pacific and Atlantic coasts as well as The Gulf of Mexico. There are over a dozen states that have coastlines directly affected by these dumpsites. If that isn't enough, of these 26 dumpsites, the whereabouts of only half of them can now be located because of poor record keeping. As a result of international chemical weapons treaties, there are currently stockpiles of chemical agents in the form of canisters and shells waiting for disposal in Alabama, Arkansas, Colorado, Indiana, Kentucky, Maryland, Oregon and Utah. The largest of these stockpiles is the Desert Chemical Depot in Utah where an estimated 6,200 tons is stored and now undergoing disposal. The most dangerous of the sites is located in Edgewood, Maryland where there are various chemical weapons, due for disposal, stored in close proximity to Edgewood area schools.

What these instances point out is the past cavalier attitude there has been concerning the management of bio and germ agents and these fiascos only involved the United States. Can you imagine the mismanagement and harm that's been done in other countries known to have chemical weapons? -Countries the likes of India, Pakistan, Russia, Egypt, Iraq and Iran. It kind of makes you hesitate when it comes to taking a deep breath.

# Part Three: Did They Get You?

Now, for some more interesting facts that are enough to make any villainous aliens green with envy - unless, of course, they're green to begin with.

In 1931 an experiment was carried out in Puerto Rico by the Rockefeller Institute for Medical Investigations. In essence, doctors infected unknowing subjects with cancer cells which resulted in 13 deaths. The same doctor who headed this project went on to become the head of the U.S. Army Biological Weapons Division and served on the Atomic Energy Commission where he continued his experimenting on America soldiers, civilians and hospital patients - this time exposing them to fatal doses of radiation.

In 1942, as part of the Manhattan Project, 4.7 microns of plutonium were injected into soldiers at the Oak Ridge, Tennessee facility that manufactured the atomic bomb. This was done without consent and without the soldiers or experimenters knowing what the results would be. These same types of experiments were also carried out at the University of Rochester and the University of Chicago with various doses of uranium and plutonium ranging from 6.4 to 70.7 micrograms. It didn't take long for the results to become apparent as most test subjects soon died of kidney failure.

In yet another round of 1942 experiments, 4,000 "volunteers" belonging to the U.S. military along with 400 black prison inmates were injected with chemical agents and malaria microbes respectively. Neither of these groups was told of the risks involved and again, many deaths were the result.

Remember that question I asked awhile back? – How would you test a bio or germ agent to verify its potency and potential? Well I don't know how you might answer it, but I would probably say, -on mice, maybe -or perhaps monkeys could offer a means for verification purposes. Now I know many who read this are strictly against any animal testing or experiments on animals of any sort. It so happens I am too. However, to my way of thinking, the greater good must prevail and being human (arguably, some might think) I tend to cheer for my side in the battle against sickness and suffering. Humanity comes first when attempting to find possible cures for infectious diseases. I, of course don't want to see any animal suffer, but I don't want to see any human suffer even more. If humans absolutely must be used in experiments to test unknown factors related to cures, I believe it must be done with volunteers who are told all the risks and unknowns related to the experiments. Perhaps death row inmates who somehow want to "redeem" themselves by making a contribution to society could be asked to make these sacrifices, but only after they are made aware of the potential dangers involved and are assured they would be treated humanely. In any event, people need to come first and any subjects, be they animal or human, should be treated with care, kindness, dignity and respect.

Under no circumstances should any experimental substances be forced upon any unknowing human being, especially innocent civilian citizens, the sick, the elderly, infants and children. Unfortunately, in the early 1950s, the U.S. military didn't see it that way.

From 1950 through 1953, the U.S. Army released chemical and biological agents over several U.S. cities and one Canadian city. These agents included disease producing bacteria and viruses that caused respiratory illnesses to thousands of people who happened to be in one of those cities at the time. The cities affected were: Fort Wayne, Indiana, Leesburg, Virginia, Minneapolis, Minnesota, the Monocracy River Valley in Maryland, St. Louis, Missouri and Winnipeg, Ontario, Canada. This was only the start of a massive program by the military to test a variety of chemicals and viruses on unknowing, American and Canadian men, women, children and infants. Many of the experiments on these civilian populations involved highly toxic Zinc Cadmium Sulfide Gas that can be fatal if inhaled or ingested. The military continued these types of tests well into the 1960s.

It seems children have been a favorite target for researchers as was the case in 1953 when 41 children, ranging in age from 8 to 14 years, had their abdomens purposely blistered with cantharides. The study, reported by *Clinical Science* was to determine how severely the substance irritates the skin. The children all lived, but throughout their lives, the physical and emotional scars from the experiment have taken their toll.

One of the most startling events occurred in 1950 when the U.S. Navy sprayed the city of San Francisco with clouds of Serratia Marcescens, a bacteriological agent and Bacillus Globigii (BG) bacteria. They wanted to monitor how these agents dissipated over a large city. Measuring devices, explained as part of a simple air purity monitoring test to improve the city's air quality, were secretly set up throughout San Francisco to record the higher bacteria levels. Along with the data collected were reports of significantly more flu symptoms among citizens and the death of one unfortunate individual who had a low tolerance to the bio agents. Over 8,000 people reported pneumonia-like symptoms to their doctor. Eventually the illness spread to surrounding communities whose citizens also reported the same symptoms which in many cases lasted for weeks. In all, 800,000 people were affected. The Navy termed the test a resounding success.

Avon Park, Florida, Hampton, Newport News and Norfolk, Virginia, Savanna, Georgia and Tampa Bay, Florida became targets of repeated U.S. Army bio weapons tests throughout the 1950s and 1960s. In one test, on Savanna and Avon Park, millions of mosquitoes infected with yellow fever and dengue were released with devastating results. Hundreds of residents suffered as they became ill with fevers, respiratory distress, encephalitis, and typhoid. Stillbirths climbed to extreme levels and many deaths occurred as a direct result of the experiments these cities had to unwittingly endure. There is evidence that hundreds of similar tests were conducted all over the United States.

In 1953 the U.S. Atomic Energy Commission (AEC) conducted radiological experiments at the University of Iowa, University of Nebraska College of Medicine, and at the University of Tennessee on pregnant women and newborn infants (less than 36 hours old). The women were given 100 to 200 microcuries of iodine-131 which the AEC knew would kill some of the fetuses. The test was done to learn at what point and to what extent radioactive iodine crosses the placental barrier. The other tests conducted on 25 newborn infants weighing between 5.5 and 8.5 pounds, consisted of injecting approximately 60 rads of iodine-131 directly into the muscles of the babies, then measuring the amount of radiological damage that was done to the infant's thyroid glands. In addition, at Detroit's Harper Hospital, 65 infants and premature babies were given oral doses of iodine-131. The results of this test are still kept secret.

In other AEC tests, 500,000 acres near the towns of Hanford, White bluff and Richland, Washington, were contaminated with radiodine-131 and xenon-133 as part of what was called Operation Green Run, back in 1953. These test areas will remain unusable for years to come.

Do you notice a pattern here? A similarity with what the aliens are up to? We have conducted totally outrageous medical experiments. They have conducted totally outrageous medical experiments.

We have done this without the permission of the test subjects. They are doing so without the permission of the test subjects. We showed a disregard for human dignity. They show a disregard for human dignity. These patterns support the theory that the aliens do have an ancestral connection to us and they may indeed be ourselves at a future point in time and/or spatial dimension. The other point here is man's inhumanity to others which really makes the alien's job of genetically controlling us more justifiable. One situation that came about as a result of U.S. medical experiments that took place prior to World War II was the Nazi clam that such medical procedures were common and U.S. examples were actually used by the Nazis at their war trials as a means of defending what they did in their experiments on Jews and Gypsies. It didn't work for them and I don't think we'll be able to plead our way out of the fate that awaits us either. What we did to ourselves in the past makes what the aliens are doing to us now a simple solution they can implement without one iota of guilt. But wait - there's more.

In 1955 through 1957, U.S. researchers took advantage of the language barrier they had with the Athapascan Indians in Alaska and gave them multiple doses of iodine-131 to see how the thyroid gland reacts to the chemical in cold weather. 17 Athapascans and 85 Eskimos took part in the experiment done under the pretext of a physical examination. The test subjects were never told the real purpose of the tests. As a result, Athapascan

Indians and Eskimos in that region suffer from thyroid conditions even today. As a matter of fact, many American Indians today have an unusually high amount of thyroid problems. I wonder why? Alaska, because of its isolation from the rest of the country, continued to be a good proving ground for other tests such as "Project Chariot" in 1958 when the Atomic Energy Commission dropped radioactive materials over Point Hope. The test results are still classified.

About that same time, the AEC, in conjunction with the U.S. military set "Operation Plumbbob" in motion at the Nevada test site, just 65 miles northeast of Las Vegas. This test involved 29 separate nuclear detonations and is responsible for at least 32,000 cases of thyroid cancer among unsuspecting civilians living in the Las Vegas area. This same series of tests also resulted in 18,000 U.S. troops being exposed to high doses of radiation. Their fate is still being suppressed by the government.

We've all heard of the many underground nuclear tests that took place at the Nevada Test Site - Yucca Flats Test Range - in the 1950s and 1960s. There were 839 of them in all. But, did you know about the AEC's other underground Atomic Bomb Blasts in places like Carlsbad, New Mexico, Farmington, New Mexico and Garfield County, Colorado? From 1958 until 1973 the AEC also had plans to detonate multiple nuclear bombs of 30 to 40 kilotons in Arizona, California's Sacramento Valley, The Mojave Desert, Alaska and Mississippi. It was all part of "Operation Plowshare", a U.S. Government program to test the peaceful applications for nuclear bombs. Really - I'm not making this stuff up. The result of this 770 million dollar foray has been hundreds of square miles of radioactive laden land that is a potential threat to thousands of people living in these rural U.S. areas. Contaminates produced by the blasts include highly radioactive Tritium, Krypton-85 and Carbon-14. These cancer causing particulates are still seeping into underground aquifers, wells and streams in spite of numerous efforts to clean them up. My suggestion: If you visit any of these places, don't drink the water. The population of Garfield County, Colorado now has further concerns as Noble Energy Company wants to drill for natural gas as close as one half mile from the blast sites. This would potentially release radioactive materials into the breathable environment that have been laying underground for the last forty years. Experts agree any natural gas derived from the area would be so contaminated with radiation that it would be useless. By the way, during that same time frame the Soviet Union had its own program that paralleled "Operation Plowshare". They called it "Nuclear Explosions for the National Economy". I can only imagine the terrific (or should that be horrific) consequences that came out of that.

Now, if you think what happened in lightly populated areas was unique, let me tell you what happened in New York City. Truly bizarre is the only way to describe the incident that took place in

New York City in 1966 when soldiers, dressed as civilians and armed with bacteria filled light bulbs, invaded the New York subway system. After casually staking out their surroundings, they proceeded to break the light bulbs by hurling them against ventilation grates and onto the subway tracks. The bacteria, Bacillus Subtilis Variant, soon spread throughout the subway system as speeding trains unknowingly helped in its distribution. The test was done to see how fast and to what degree a germ or bio agent would spread and then dissipate in the subway environment. It didn't take long for them to find out. As they monitored the results, the exposure produced widespread flu symptoms among many New Yorkers, but at the time no one knew the real cause of the outbreak. More than one million civilians were exposed during the experiment. Although no one died as a result, it is still one more example of the military testing biological weapons on their own civilians. Were you in New York in June 1966? If you were and you came down with the flu, you now know why.

In another 1966 incident, this time occurring outside North America, a very infamous nuclear accident took place when in January of that year, a U.S. B-52 bomber collided with a KC-135 tanker while refueling over the picturesque fishing village of Palomares, Spain. Unfortunately seven of 11 crewmen lost their lives in the collision. Just as unfortunate was the B-52's release of four hydrogen bombs. One of the bombs splashed into the coastal waters of the Mediterranean Sea and was recovered intact after three months of searching by the U.S. military. Another bomb landed without exploding just outside the village, but the high explosive igniters on the remaining two bombs detonated on impact, spreading a deadly mix of radioactive material including plutonium over a vast area of the Spanish countryside. Although these explosions did not set off a nuclear reaction, the cleanup effort following the accident was immense as hundreds of tons of contaminated soil were removed from the 25-acre site and (take note Nevada) shipped to the United States for disposal. In all, 182 million dollars for the recovery, clean up and related claims were paid by the U.S. as a result of this massive nuclear nightmare. But, it doesn't stop there because along with the Palomares fishing industry, this area of Spain happens to be a source for one other product consumers of fine food delight in dinning on. That item is snails. These relatively small creatures are devoured in droves throughout the world, but mostly in Europe and Asia where they are considered as fine a delicacy as there is. Now, a frightening problem has risen as the snails of Palomares, that have been exported all over the world as well as eaten by the locals for the entire 44 year period following the accident, have been found to be highly contaminated with radioactive particulates from soil that was somehow overlooked during clean up operations at the crash site. An even more immense clean up is now planned for Palomares. Escargots anyone?

As a point of interest, the U.S. Air Force admits to 32 lost nuclear bombs or "Broken Arrows" as they called them. These losses all occurred during the cold war with the Soviet Union and most have been recovered from both land and sea locations. The environmental group, Greenpeace estimates, during that same period, 50 nuclear warheads were "lost at sea" and now, after 40 to 50 years have passed, still lay deteriorating on the bottom of the world's oceans and lakes. Of these 50 warheads, the great majority are believed to be Soviet. This information, first told to me by the astronauts, is backed up by the Brookings Institute, a Washington, D.C. think tank. When you combine this information with the accounts of nuclear waste plowed into the South Pacific from the failed tests done there in the 1960s, our planet starts looking like one giant nuclear dump. When most of these accidents occurred, scientists believed plutonium and uranium dissipated in water over time. To the contrary, they now find these radioactive substances actually stay active in water for extremely long periods of time because they cling to natural mineral deposits such as colloids and can travel, attached to these deposits for long distances, contaminating most if the world's water resources as they go. Now how about a nice glass of water to go along with those escargots?

In still another debacle that points out how man's technological prowess has outpaced his common sense, there is this bit of information: In 1970 The Department of Defense (DOD), under CIA supervision and with congressional funding, started developing a new synthetic Biological agent at the U.S. Army's top secret bio weapons facility at Fort Detrick's Special Operations Division. It seems the DOD wanted to create a disease that would be resistant to all known immunological processes and therapeutic treatments. They wanted a killing agent that had the potential to wipe out whole continents of humans in a relatively short time. This new agent had to have no cure to impede its deadly progress. With an initial 10 million dollars for research and development, it took researchers just five years to come up with a prototype of the virus. Meanwhile, in an unrelated program at the Center for Disease Control (CDC) in Atlanta, which is another government-funded facility, a new vaccine was being tested for the treatment of Hepatitis B. As it turns out, Hepatitis B was especially prevalent in gay men. Consequently, gay men were chosen to take part in the Hepatitis treatment trials at the CDC. Well, I bet you can tell what's coming next. It was in 1976 that scientists believe the AIDS virus first made its appearance. Was it simply a coincidence that the first victims of AIDS happened to be homosexual males? Was that first edition of the DOD's killer synthetic bio weapon actually what we now know as HIV, - the virus that causes AIDS? Were these gay men more unwitting civilians who became victims as they thought they were being test subjects for a Hepatitis B treatment and instead were given doses of the DOD's new killer virus in another experiment gone awry? This is how it was told to me, but as I've said before, you're invited to draw your own conclusions.

In a final example of government ineptness, I'll present you with a chilling account of what I think is one of the most outrageous of these so-called tests. This actually involved a series of tests covering several years. Dugway Proving Ground is located approximately 87 miles southwest of Salt Lake City, Utah. Dugway was the site that provided the necessary isolation and secrecy for what military scientists were doing there back in 1952. This became their operations base for determining how biological and germ agents might spread across the United States. What better way to get the answer than to actually release strains of Brucella suis and Brucella melitensis over the entire country. That's exactly what they did. Every populated area, over a multi-year time span, was affected. Not one state within the continental U.S. was spared. As a result, every person who lived in the United States during 1952 and beyond has been infected with these agents. For those born after this period, the Brucella bacterium is genetically past on from generation to generation. So, to answer the question this particular part of the book asks at the beginning; "Did they get you?" Let's just say, if they didn't, you're probably new to the U.S. and if so, - to you I say good luck but keep your eyes on the skies because those jets that are flying over just might be leaving more in their vapor trails than water droplets. Dropping anti-toxins or toxins doesn't seem to make much difference to those in charge.

# Chapter Six
# The secret U.S. nuclear meltdown

## Part One: A Recipe for Disaster

Throughout the 1960s and 1970s, residents of Southern California's San Fernando and Simi Valleys were periodically jolted, not by earthquakes that happened occasionally, but by the sometimes daily roaring of rocket engines being tested at, what was then, North American Rockwell's Rocketdyne Division. The plant, that was situated against the picturesque Santa Susana Mountains, was built during a time when the area was fairly isolated. Post WWII development soon brought affordable homes, schools, shopping malls and recreation to this once desolate region. It was during this time that the San Fernando Valley became one of the fastest growing communities in the country. New residents poured into this northwestern suburb of Los Angeles that, in spite of its sometimes stifling heat, offered affordable housing, plentiful jobs and the potential for unlimited growth. The valley became a great place to raise a family as business and educational opportunities presented the promise of a good life for all. But, the economy wasn't the only thing that was booming in the valley.

The familiar sounds of the rockets could be heard as far away as Santa Barbara to the north and Long Beach to the south. No one seemed to actually mind the noise since it represented America's hope for progress in a race with the Soviet Union to develop super rockets that could take us to the moon and beyond. What wasn't known to most residents was the fact that long before rockets were tested at the facility, it was an experimental nuclear power plant.

Known as the Santa Susana Field Laboratory, the 2,850 acre site, adjacent to the heavily populated community of Simi Valley, was opened in 1948 as an uncontained sodium reactor. This means it was considered an open field lab with no containment possibilities in the event of a nuclear accident. In reality, the lab housed 10 nuclear reactors along with plutonium and uranium carbide fabrication plants. The facility also had what was called the "Hot Lab" which was used for remotely cutting up irradiated nuclear fuel shipped in from other federal nuclear plants. The Santa Susana facility was one of the first major nuclear power plants in the United States and as such, had none of the safety features all nuclear power plants have today. The reactor building, for instance was made entirely out of thin corrugated steel panels and the main reactor itself was built with only a single wall separating it from control room engineers. It operated as a power source for the area, in tandem with the rocket engine testing, for almost 30 years in spite of the two seemingly non-compatible sciences being so unimaginably close together. By today's nuclear plant standards, the Santa Susana Field Laboratory was a crude undertaking from the very beginning. Rocketdyne set up their multi-million dollar engine testing facility at the site because it was offered to them as part of the lucrative federal contract to build badly needed engines for various missiles. At the time, it probably presented itself as a very good location for the necessary testing as well as a very good business deal. The company (now a part of Boeing) is still feeling the ramifications of that decision. With no nuclear regulations in place and no monitoring by any government authority, the Santa Susana Field Laboratory went about its business of generating nuclear power to the surrounding area without the slightest regard for public safety or for the safety of employees at the site. It became inevitable that someday time would catch up to the lab's shoddy practices.

Whatever can happen - will happen. Perhaps you've heard this as part of Murphy's Law. It never became truer than on July 13, 1959 when all hell broke loose at the Santa Susana Field Lab. It started when a clogged coolant channel in a 20-megawatt nuclear reactor caused fuel rods, used for heating water in order to produce steam, to rupture. Radioactive gases containing iodine-131 and cesium-137 were then released into the plant and subsequently into the atmosphere. The result was a 30% reactor core meltdown. This partial meltdown was the third largest in nuclear accident history, only eclipsed by Chernobyl in the Ukraine and Windscale in the U.K. The release of radioactive iodine was even worse than the infamous Three Mile Island accident that occurred 20 years later. Some estimates put the SSFL incident at 100 times worse than Three Mile Island. In trying to identify the cause leading up to the accident, investigators found the sodium reactor had a problematic amount of contamination in the core that was never addressed. As the contamination worsened, the reactor was unexplainably left on line and this is what lead to the eventual clogging of the coolant channel and the meltdown.

---

# Part Two: A Lack of Candor

Now it so happened that Rocketdyne, ie-
North American Aviation, and later -Rockwell, liked
their little rocket testing facility out there in the
valley. So they immediately proceeded to downplay,
and in some cases cover up what happened at the
SSFL. Although the official word was that a mishap
did take place, their statement went something like;
"We have concluded, no radioactive substances
were released and no one was injured." The public
accepted the company's word and the incident was
soon forgotten. That is until 20 years later. It took
that long before people started wondering why more
than 27% of Santa Susana Field Lab workers were
dying from various forms of cancer that included
cancers of the thyroid gland that was eight times
higher than the general public. Thyroid Cancer is a
direct result of exposure to iodine-131 which, as you
know by now, is a by-product of fission nuclear
reactors like those in operation at the SSFL. In
addition to employees' elevated cancer levels, there
have been reports of deaths among area residents,
also from blood and lymph node cancers, that go
well beyond the norms. The Los Angeles Times
reported on Oct. 6, 2006:

> *Radioactive emissions from a 1959 nuclear
> accident at a research lab near Simi Valley
> appear to have been much greater than
> previously suspected and could have resulted
> in hundreds of cancers in surrounding*

*communities, according to a study released Thursday (Oct. 5, 2006). Chemical contamination from rocket engine testing at the site continues to threaten soil and ground water in the area around Rocketdyne's Santa Susana Field Laboratory, the study also found.*

Oh yes, - I neglected to mention, in relation to the rocket testing, and in addition to the nuclear contamination, Perchlorate, a highly toxic component of rocket fuel, and Trichloroethylene, (TEC) a cleaning chemical that causes brain and liver cancers as well as kidney and heart damage were routinely washed into the soil and groundwater surrounding the unregulated plant. Contaminated chemical and radioactive combustibles were also disposed of by burning them in an open-air sodium burn pit in violation of restrictions prohibiting such activity. This happened over a thirty-year period beginning in 1954 and lasting throughout 1983. It is estimated that more than 500,000 gallons of TEC alone were dumped at the site. This pollution of both air and groundwater potentially caused widespread outbreaks of cancers among Field Lab workers and the public. -Back to the L.A. Times article.

*The nuclear meltdown, which remained virtually unknown to the public until 1979 could have caused between 260 and 1,800 cases of cancer, within 62 square miles surrounding the facility "over a period of many decades," the study concluded. But the advisory panel that oversaw the five year study, conducted by an independent team of scientists and health experts, said it could not offer more specifics about potential exposure to carcinogens because the Department of Energy and Rocketdyne's owner, Boeing Co., did not provide key information. "This lack of candor... makes characterization of the potential health impacts of past accidents and releases extremely difficult," the panel concluded. Boeing officials vigorously disputed the findings, saying the study was based on miscalculations and faulty information.*

A lack of candor - indeed. Strangely enough, the 1959 meltdown wasn't the only nuclear accident rumored to have taken place at the Santa Susana Field Laboratory. Throughout the operational life of the SSFL, several mishaps have purportedly taken place with the same amount of "candor" being shown by Rocketdyne. The big difference between these accidents and the July 13, 1959 incident is the secrecy that shrouds these other situations. The company has been very good at keeping these nuclear disasters top secret. One of the worst took place in 1964. Very little is known about how this one started. Not even my astronaut buddies could shed much light on exactly what transpired, but they did tell me there was an 80% meltdown of the fuel rod claddings in one of the reactors. That alone tells you the scope of whatever happened was extraordinary. This was the final snafu for the SSFL experimental nuclear power plant as the government decommissioned it one year later. The end result is now a multitude of law suits that are either taking place or pending against Rocketdyne's successor, Boeing and of course all the misery that workers, their families and residents who lived near the plant are enduring.

Was there a lesson learned? The answer is yes, but I'm afraid it has been learned too late as what happened at the Santa Susana Field Laboratory becomes just one more reason for an alien annihilation of humanity - as we now know it.

---

# Chapter Seven
# Reviewing Data

## Life Cycle Significant Items

One of the more outstanding consumers of time during Skylab checkout was the company's policy of reviewing data in a test procedure to insure no processes were overlooked and all verifiable items in a procedure were "bought off," - that is inspected and stamped for being inspected and/or verified. One aspect of this process was the traceability of "life cycle significant items." These consisted of switches, relays and circuit breakers that supposedly had a predetermined life usage linked to their possible failure rate. A switch, for instance, might be considered to have a 99.9% chance of not failing during its first 20 cycles (on and off). After that it might only have a 95% chance of not failing for its next 20 cycles. After that, its estimated failure rate increases exponentially to only a 70% chance of not failing for its next 20 cycles, and so on. The point is, this documentation involved keeping track of every time a switch, relay or circuit breaker was thrown. When these items reached a certain percentage, say a 30% chance of failing, they had to be replaced by a new component. With hundreds of these components on board and each test procedure cycling them either manually or automatically by computers, the job of replacing them became a tedious nightmare. This is what I was reminded of as I started to review and arrange my notes for this book. At the time, I also began to wonder -Who's going to actually believe these things that even I was still astonished at reading? How could I ask people to seriously accept the premise that unknown extraterrestrial forces are out there, scurrying about the galaxy aboard asteroids that they've converted into their own personal hot rods, while sending their android workers down to Earth to change us into a race akin to zombies so they can eventually take over the planet and steal all of our heretofore mismanaged natural resources? Oh yes, that sounds reasonable. I put my notes away and decided I really didn't need the aggravation this book would surely bring my way. Then on 9-11-2001, my trepidation ceased. It was then that I knew for sure the truth must prevail as it became obvious that man's technological prowess is only eclipsed by his social ignorance and his ability to destroy. I realized it doesn't matter if anyone believes this book or not. It is nothing more than what I was told, in confidence, by a group of extremely well educated and highly trained individuals who definitely were in positions to know what the truth is. I know the serious manner in which the subjects discussed in the book were presented to me and I know what is told here is all true. Again, I'll leave the believing up to each individual reader and if in the future, the matters discussed herein prove to be either true or false, so be it. I have merely reported to you, what was told to me.

It is certainly reasonable to assume that there are those who hold powerful positions within world governments and controlling organizations who also know the truth. Let it be said now that if any mysterious acts of wrong doing befall me, it will come from these entities that do not want the truth to be known. Any act against me or to those close to me, from here on, will only help prove the truths, as told in this book, are valid.

I spread the notes out before me like a giant fan. Some were dated, but others were not. Some were scribbled and some printed neatly. None of them were in any kind of order. As I rummaged through them, memories popped in and out of my mind - such as ...This fellow sure talked a lot. Who told me that? What was he referring to here? This guy hardly said anything. Why in the world did he tell me that? Memories…memories. I started to sort the notes into cohesive piles. I can't answer for the astronauts but I really think they believed I would never divulge what was said, and - as I had been told - if I did tell anyone else, no one would believe it anyway. I can only imagine the secrets they did keep to themselves. So I plowed forward with this book project knowing I would disturb some people, but as knowledge is power and truth will always prevail, I'm hopeful I've created a manual for the greater good.

This book, without a doubt, encroaches on certain security restrictions. I do believe however, that the people have the right to know the possibilities that lie in their own future. My sole purpose for the book's existence is to bring you the hidden truths that will affect your life and your future decisions. I have endeavored to do that along with the how and why of those truths. I don't know why these accounts found their way to me. I certainly didn't ask for this role as a reporter of doom. The astronauts of Skylab didn't have to tell me anything, but they did. So too, I must let you know what I know and let the consequences become my fate. If you doubt what is written here, I don't blame you a bit. But I will tell you, my discussions with the astronauts about these episodes were as serious as it gets and their accounts are proving to be more and more valid every day. I now see evidence of the stories they proclaimed popping up all over the place, from newspapers to the internet. The overwhelming conclusion is, there is just too much evidence and too many witnesses to chalk it up to coincidence alone. My hope is to get this message out before we are all lost in a sea of ineptitude. Just as the Titanic started listing to its side, I can feel the slightest amount of list starting for the entire human race. A faint bit of off balance here and a bit of stumbling there. Slowly our demise is drawing nigh. Without pain, without suffering we will return from which we came. A more natural world awaits us and our posterity. Perhaps, in our new surroundings our talents for getting along with each other will not become generations behind our technological development as it has in this era. Perhaps we will learn to live with one another without cruelty, jealousy or revenge on our minds. Perhaps conquering sickness and disease, without conquering each other, will become our highest goals. Perhaps we will truly consider all people as being equal and the wealthy will be those who are rich in spirit and knowledge and willing to share that wealth with everyone. Perhaps all of this will happen soon, perhaps much later. I do not know the precise timing of any future event but the following might give us a hint. I'm talking about an asteroid that is hurtling headlong toward the Earth and is the size of a small mountain. This monster is called 99942 Apophis. Now it seems Apophis, a.k.a. 2004 MN4, which was detected in 2004, will pass by the Earth in the year 2029, barely missing us by a mere 23,000 miles. This 1,000 foot diameter chunk of rock will be traveling at about 15 miles per second and actually dip below the orbits of our communications satellites. Could this be our alien friends' final approach? I tend to think the alien commanders on board Apophis want to take a closer look at us for themselves before they follow through with the final phase of their plan to destroy all human intellect. Apophis is ironically (and appropriately) named after the ancient Egyptian god Apep, "the destroyer" who dwells in the eternal darkness of the underworld. After this god of the underworld completes its "flyby," scientists calculate it will enter a half mile wide region of space, nicknamed "the keyhole." If the asteroid passes through the keyhole it will, without a doubt, return to slam into the Earth on April 13, 2036. You can take it from me, this won't happen. As I've mentioned before, the aliens want our planet in topnotch condition and if this asteroid were to hit the Earth, it would produce 100,000 times the energy of the atomic bomb dropped on Hiroshima. Since the aliens have the technology to control asteroid trajectories, Apophis, - or any other truly large object from space, alien manned or not, will in all likelihood never hit the Earth… at least not in the foreseeable future. Their plan for us is much more subtle than throwing rocks at us and allowing our ecosystem to be destroyed.

They have a new world in mind for us and our antics of mismanagement, disrespect and disregard for our own planet Earth. This, along with their need for our natural and/or human resources is why their plan has been set in motion. I was told by the Skylab astronauts that they are using genetic manipulation to rapidly induce Autism, Alzheimer's and other brain disorders into our society. I was told, back in 1972, these anomalies would increase exponentially throughout the coming decades, and they have. I was told factual sightings of UFOs would decrease in time as alien invisibility and stealth technologies improved, and I believe that is happening. I was told of many outlandishly poor decisions made by humans in the past that are now taking their toll on the Earth, and it has all proven true. I was told to look for a period of mediocre accomplishments in our education system and a slowdown in our non-computer designed inventiveness. This also is taking place. Most of all, I was told a future manned trip to an alien occupied asteroid, to confront these entities, could become a reality and that is now in the planning stage at NASA. I, in fact believe the whole Orion Moon / Mars Program is nothing more than a cover story and prelude for the asteroid trip. Everything I was told and jotted down in notes, for the entire two years I worked on Skylab's check-out team is now beginning to fall into place. Like pieces of a puzzle, each and every part has lined up and interlocked to verify the astounding testimonies I learned during that time. I honestly don't know how the story of humanity will end, but as one astronaut speculated:

*"I believe there is hope for a better tomorrow. Although, in the short term things might look bleak, in the long term, I believe all will turn out for the betterment of mankind. Despite the fact that we have unquestionably messed things up this time around, in the end, the true irony will bring us back to where we started but with new knowledge and understanding about how to treat each other- and the planet - in a much better way."*

After putting the pieces I was given to this puzzle together, I now agree with the theory that this one astronaut brought to my attention. I believe these so called aliens are none other than ourselves from some future time, coming back at this crucial juncture in history, in order to do what is right for the human race. They, the aliens who are actually our posterity, will, with compassion and care, like parents with a newborn infant, put us to bed with the hope of an enlightened future, for both us and for them. Whatever road the story of mankind takes, it will not end now or even in a millennium. We as human beings will prevail in the future as we have in the past. We are now, in fact, taking our first designed steps at exploring asteroids - perhaps our future homes. Will the knowledge we gain from this now, somehow spark our future imaginations and cause us to use asteroids for space travel in a far off time and place?

We are the ultimate of life cycle significant items and we now find ourselves being replaced as our failure rate has reached a critical point. Whatever those whom I have called aliens gain from our replacement, be it Earth's resources or their own existence, I believe our sacrifice will be remembered with great gratitude by an unexplainable universal life force in an inconceivable future and although all our future dreams may die, the remembrance of our being shall never die.

Stay vigilant and aware… as long as you can.

# # #

# Epilogue

(Or, random thoughts from my notes that just don't seem to fit anywhere else in this book.)

So here we are, sitting in our comfortable easy chairs while this giant rock we call Earth spins through the universe at breakneck speed. In our daily pursuit to gain happiness and make financial hay we hardly have time to think about what the whole shebang is all about. Just what is this trip we all find ourselves on? It's remarkable we can even contemplate our place in the cosmos at all. When we do think about the philosophical questions concerning the universe, we are left with a carload of even more questions that only lead us to realize just how mysterious our surroundings are. How can we possibly begin to know anything when our abilities to sense true reality are clouded by things like dark energy, dark matter and the lack of an all encompassing theory to explain all there is - not to mention the constant alien attack on our mental competence. The big question here is: Does - what we perceive as reality - really exist at all? This subject came up often in my discussions with the astronauts and I'll tell you what they had to say. I was told: Reality, as we perceive it, does not exist. What your senses are picking up and distributing to your brain is only a reaction to cause and effect. What your conscious mind perceives as reality is a pseudo reality, made from the afterglow of the past and a precursor of the future. The thing we call "the present" is non-existent in the space/time element we are seemingly part of. In actuality, all that is (or ever will be) is happening at the same time and place in space/time. Physicists call this the Many-Worlds Interpretation. It is linked to the quantum physics phenomena called entanglement which is predicted by Einstein's Theory of Special Relativity. A good test of this non-reality phenomenon is to try to envision the past happening now, in the present. It's impossible, right? What's past is gone from our perception of reality. Now try to envision the future happening now, in the present. Again, it's impossible because the future has not yet taken place according to our perception of reality. Now try to envision the present happening now in the present - whoops, sorry the present is already in the past, gone, out of sight and possibly billions of light years away. Think of a high speed train rushing toward you. It is the "time train" and it never stops. It blasts past you leaving cinders and dust in your eyes then disappears into the night. That is how the future and past work. The future is coming at you faster than the speed of light and the past is moving away from you at the same speed. There is no present and thus no reality. Do you really think the train would stop for you? So we find ourselves suspended in this no-man's land we call "the present". We think we're in a stable place where everything seems normal but our science and mathematics tell us reality - really does not exist. Here-in lies the paradox. If we don't have reality to bounce around in, what do we have?

Cosmologists, throughout the world, are now employing computer programs that simulate the formation of the universe. These simulations can bring entire virtual universes into existence with just a few keystrokes. In fact, many separate and very different universes can be created by these programs. Researchers can bring about situations where the atomic structures that existed during the formation of these virtual universes can be tweaked so the end result could be very unlike the universe we see around us. This is another way to examine the "Many Worlds Interpretation". Now here's where I would like you to simply relax your mind and consider the following: If throughout this inter-dimensional, time transcending multi-verse there are hundreds or thousands of virtual computer programs spitting out hundreds of thousands or millions of simulated universes, what do you suppose the chances are that the one we find ourselves in is the one and only real one? Reality, by definition, means a non-ethereal, positively actual situation. If there is only one real reality, imposters, no matter how detailed, cannot be it. It is truly a million to one shot that our universe actually exists. What our senses perceive as reality, in all likelihood, is nothing more than a virtual place in a virtual time created by some far off computer program in a far off, but very real universe. Perhaps there is a reason we see this all as we do.

Through the use scanning electron microscopes, scientists have detected strange looking appendages that cover the surface of the human brain. It is believed these tiny "micro-tubules" act like antenna searching for a signal. Could it be - they are receiving information from outside our perceived environment? The brain has many functions. Most involve our nervous and sensory systems. The brain controls all the functions of our bodies like breathing, heart rhythm, movement, speech, etc. But how do we come up with abstract ideas? Things we have never contemplated before seem to pop into our thoughts all the time. Things we never knew are suddenly known without the slightest idea of how we thought of them. To some, outstanding intuitive abilities are considered normal. Savants and child prodigies surprise us with their special talents. Where does it come from? We can see with eyes closed, hear, taste, touch and smell things only with our minds. We need no senses to feel what it's like to ride a roller coaster. We can go to the moon and back in a flash. Twirl around the galaxy or taste a cookie all at the same time. Can any of this be possible with only our basically primitive brain, the controller of bodily functions? Or, is there more to this universe than we can detect? The mind seems to truly be separate from the brain. Every astronaut I talked with on this subject agreed, our mortal selves are somehow in touch with our spiritual immortal selves and the transmissions (for lack of a better word) our brains are receiving are completely controlling our thought processes. These transmissions are emanating from a sphere beyond our understanding and this realm is where we truly reside. It is where the past, present and future are all melded into one and time has no affect. It is neither a place nor a time. It is eternal and thus, the only true reality. My astronaut friends said this was never so evident than when they were isolated in space. For it was then that they realized; Our bodies and brains may perish but our true selves, what we perceive as our minds, will always live on in a far more unrestrictive and exquisite existence than we can ever now imagine.

The transition is at hand.

---

I would like to thank the following individuals and organizations for their inspiration in bringing this book to completion.

United States Army:
Maj. General, David B. Lacquement, Army Intelligence & Security Command.
Lt. General, John F. Mulholland Jr., Army Special Operations Command.

United States Air Force:
Lt. Gereral, Raymond E. Johns Jr., Strategic Programs.
Maj. General, C. Donald Alston, Nuclear Integration Office.
Col. Michael G. Caldwell, Public Affairs.

NASA: (National Aeronautics and Space Adm.)
Robert Jacobs, Assistant Administrator - Public Affairs.
James B. Garvin, Former Chief Scientist.

MUFON: (Mutual UFO Network)
James Carrion, International Director.

DIA: (Defense Intelligence Agency)
Lt. General, Ronald Burgess, Director.

...And, of course, all the Astronauts, Engineers, and Technicians who worked on Skylab.

*Death of a Trillion Dreams*